A GEOCHEMICAL ATLAS OF THE PORTUGUESE MINERAL WATERS

Front page: The spring "Facha" in Vilarelho da Raia (district of Vila Real).
Photograph taken by José Manuel Marques.

A geochemical atlas of the Portuguese mineral waters
by HGM Eggenkamp, JM Marques and O Neves

©2015, The authors

ISBN 978-90-816059-6-0 (Soft cover; CreateSpace)
Onderzoek en Beleving, Bussum, The Netherlands
post@onderzoek-en-beleving.nl

H.G.M. Eggenkamp
J.M. Marques
O. Neves

Centro de Recursos Naturais e Ambiente (CERENA), Instituto Superior Técnico, Universidade de Lisboa, Av. Rovisco Pais, 1049-001, Lisboa, Portugal

A GEOCHEMICAL ATLAS OF THE PORTUGUESE MINERAL WATERS

PREFACE

It is a great pleasure for me to write some words about this Geochemical Atlas of the Portuguese mineral waters.

First of all because, since the seventies of the 20th century, I have launched and promoted at Instituto Superior Técnico of the University of Lisbon (in the framework of Centro de Petrologia e Geoquímica - CEPGIST) a research project on mineral and thermal waters, balneotheraphy and geothermics were the authors of this study have been research workers.

Secondly for our interest either as President of the National Council for Thermalism (Comissão Nacional de Termalismo) where the first general approach was made to correlate the geothermochemistry of the Portuguese mineral waters with their effect on human health (therapeutic applications), or in the field of low enthalpy geothermal applications of the Portuguese thermal waters.

Finally we have now a quite nice and good presentation of an outlook of a geochemical atlas of the Portuguese mineral waters using the chemical data available in the literature.

Fifteen geochemical maps have been prepared, thirteen of these including chemical elements and compounds and another two on TDS and pH. These maps are taking into account about five hundred chemical analysis available in the literature, and show the close relationship between the geothermochemial signatures of the mineral and thermal waters and the mineropetrography of the geologic substructum and the geotectonics of the areas from where these waters are issued. These maps are also quite elucidative and useful for several scientific and geochemical purposes (like geological, chemical, medical and environmental sciences, and also technological applications in industry, balneotherapy and geothermics).

I hope that within a short time new steps could be accomplished regarding future studies involving chemical analysis of minor and trace elements which would enlarge the application field of a new geochemical atlas either in environmental geochemistry or in medical geochemical domains.

Luís Aires-Barros
President of the Academy of Sciences of Lisbon

SUMMARY

In order to understand the distribution of mineral water types in the Portuguese Mainland geochemical maps of Li^+, Na^+, Mg^{2+}, Ca^{2+}, Al^{3+}, Fe(T), F^-, Cl^-, SO_4^{2-}, NO_3^-, ΣCO_3, HS^- and SiO_2, as well as the total dissolved solids and pH were prepared. These maps were prepared using concentrations of these chemical species in mineral and thermal waters. The purpose of this study was to asses whether viable geochemical maps could be prepared using the chemical composition of mineral and thermal waters. All analyses were taken from open literature sources. The advantage of this approach is that no field work campaign had to be set up. The study was composed of over 500 mineral and thermal waters ("sample points") with an average density over the country of one sample point per 100 to 300 km^2, where the density in the north of the country was about three times as high as in the south. The maps showing the distribution of these chemical species could be related very well to both the geology and the geochemistry of the country. An example of this effect is the distribution of the total dissolved solids. In regions with evaporite deposits in sedimentary basins in the south and west of the country the mineralization of the thermal and mineral waters is high, while in areas in the north of the country which are represented by extensive granite intrusions very dilute waters are most commonly found. Also most other species for which we were able to produce geochemical maps follow a distribution that reflects the geochemical signatures of the rocks through which the waters percolate properly. An exception is the distribution of nitrate, which in many cases reaches high concentrations due to fertilization of the land so that it is more representative for agricultural contamination. The risk that a mineral water is contaminated is especially high in regions where mineral waters have a relatively shallow migration path. Compared to other geochemical mapping techniques, such as those based on stream sediment or stream water samples, geochemical maps based on mineral and thermal waters generally represent the chemistry of a much deeper part of the upper crust. In some areas they even show extremely deep chemical signatures, such as the occurrence of mantle CO_2 in mineral waters in the northern part of the country. A small disadvantage of this technique may be the uneven distribution of thermal and mineral waters in a study area. However, as Portugal is exceptionally rich in mineral and thermal waters this approach does work very well in this country.

CONTENTS

INTRODUCTION

During the last decades interest in geochemical mapping projects has been increased considerably. Special attention for geochemical mapping originally developed due to regional surveys for thermal and mineral exploration but more recently (since about 1975) it has moved towards environmentally applied geochemical studies (GARRETT *et al.* 2008). Geochemical Atlases that were produced in these studies contain maps that show the distribution of chemical elements from specific media such as soils, stream sediments or shallow ground waters of a certain area (such as a region, country or even continent). Modern maps are also produced to update knowledge on source, transport and fate of elements in the biogeochemical cycle as, for example, described by REIMANN *et al.* (2001). GARRETT *et al.* (2008) describe the development of geochemical mapping from plain geochemical prospecting to international mapping covering whole continents and show from which areas Geochemical Atlases already do exist. From this summary it is clear that Geochemical Atlases exist from many parts of the world, although especially areas such as central and northern Europe and China are very well represented.

Geochemical maps can take many different forms. Depending on the survey area the density of the sampling points can vary enormously. Current geochemical maps are produced with a range from high sample densities (less than 1 sample per 10 km^2) for mineral exploration and later also for environmental purposes (*e.g.* HAWKES 1976; RAPANT *et al.* 2009), to very low sample densities (like 1 sample per 2500 or even 10000 km^2) for national and international geochemical mapping such as the "Geochemical Atlas of Europe" (SALMINEN *et al.* 2005). In this European low-density (1 sample per 4700 km^2) survey samples were collected from environments such as flood plains and stream sediments, stream water and different types of soils (organic top layer, minerogenic top and subsoil).

Results of these low density mapping surveys show a good overview of the geochemical background over large areas and, as shown by SMITH & REIMANN (2008), have a robust geochemical pattern (i.e. if the same area is resampled and remapped, comparable geochemical patters will emerge). However, when more detailed information on larger scales is needed, the sampling density needs to be increased to a level of one sample per 100 to 200 km^2. As geochemical mapping generally requires a large investment in sampling (field campaigns), sample preparation and analyses, which is something that needs to be overcome when setting up a program, it might be useful to start making this kind of studies with data available from the literature. An example of data regularly available in the literature for this use is data from mineral and thermal waters. These naturally occurring spring waters have been analysed since the 19th century as they are used for medical purposes in many countries for hundreds to thousands of years. A reason for using mineral and thermal water analyses is their geochemical character. Unlike stream waters, which obtain their chemical composition from the surface geology, mineral and thermal waters obtain their chemical characteristics from deeper geological formations due to their deeper circulation. It has been shown in many studies that groundwater presents different geochemical characteristics depending on the geology they are extracted from. For example, in studies from Portugal (GALEGO FERNANDES *et al.* 2008) and the Netherlands Antilles (VAN SAMBEEK *et al.* 2000), where groundwaters were sampled from relatively shallow wells, statistically significant differences were shown between wells drilled in different lithologies, which were sedimentary in the Portuguese and magmatic in the Netherlands Antilles case studies. In these studies relatively small areas were sampled and studied. However, when collecting mineral water analyses from larger, countrywide, areas it is expected that variations due to the geology could also be detected and, using this feature, geochemical variations on a smaller scale could be studied as well.

THERMAL AND MINERAL WATERS

Thermal and mineral waters can be defined as natural (spring) waters that show a distinct temperature and/or chemistry when compared to other "*normal*" groundwaters from the area or region where they emerge, and which chemical composition and temperature are not varying over longer time scales. Within this group thermal waters can be defined as waters with a temperature above the average annual air temperature, which in Portugal conveniently is chosen as 20 °C (*e.g.* ALBU *et al.* 1997). This definition indicates that mineral and thermal waters must have a relatively deep circulation. Thermal and mineral waters normally are issued along faults and have a recharge area which is in most cases several kilometres away from the discharge location. The water gets its chemical composition as a result of water-rock interaction, along its deep underground flow path, between the recharge area and the discharge location. Thermal waters are usually highly mineralized waters with issue temperatures above 20 °C, explained by a rapid up flow from the reservoir to the surface, favoured by the presence of deep faults. On the other hand, mineral waters are also often highly mineralized, but since they usually issue along rock fractures and discontinuities, they present temperatures lower than 20 °C. Most of them are (or were in the past) also used in local Spas. A detailed description of the chemical evolution of groundwaters from rainwater to the sometimes very interesting watertypes can be found in ALBU *et al.* (1997). Using stable isotope ratios of hydrogen and oxygen AIRES-BARROS *et al.* (1995) showed that the original infiltrating water, in the recharge areas, is rainwater which

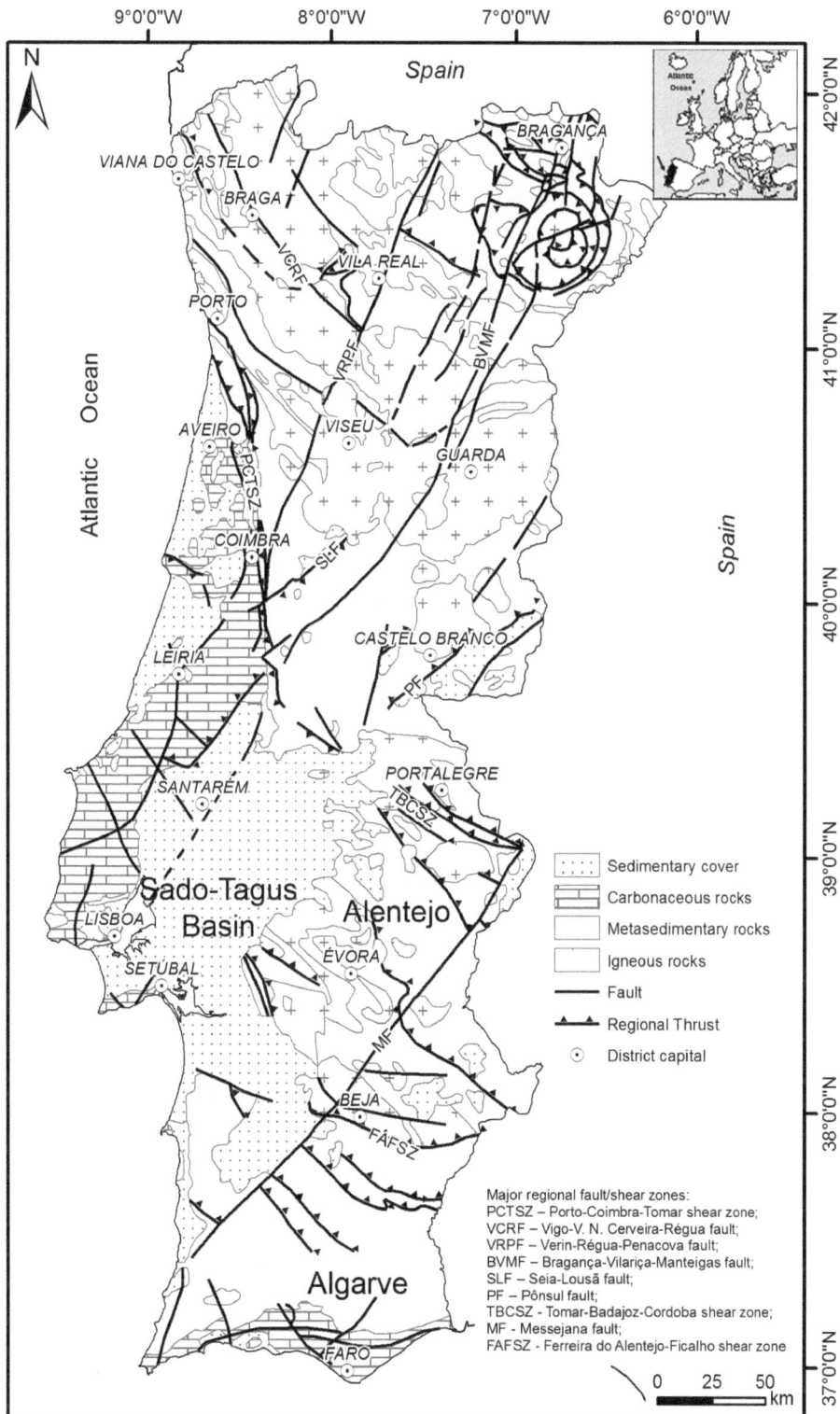

Adapted and simplified from: Geological map of Portugal, scale 1/1.000.000, L.N,E.G., 2010

Map 1: Simplified geologic map of the Portuguese Mainland. Adapted from L.N.E.G. (2010). The Western and Southern Meso-Cenozoic Basins are characterised on the map by the Carbonaceous rocks, the Iberian Massif by the Metasedimentary and Igneous rocks.

3

of course contains only very little dissolved solids, so that all mineralisation must be picked up due to water-rock interaction; in some cases from different geological formations (*e.g.* MARQUES *et al.* 2003; 2006; 2008; 2010). Using this technique, these authors also showed that the recharge areas of many Portuguese mineral and thermal waters (*e.g.* Caldas do Moledo, Chaves, Cabeço de Vide, Caldas da Rainha) is rarely further away than 10 km from the thermal/mineral springs/boreholes, which was assumed in earlier times (see *e.g.* CARNEIRO 1993). As a consequence thermal and mineral waters can be considered as representative for water-rock interaction of only a relatively small area close to the springs, and it could potentially be used as a sample point for geochemical mapping. A potential drawback may be that thermal and mineral waters are normally not very evenly distributed over a larger area. Their density is normally highest in more mountainous areas, and even there only along major faults, while in large sedimentary basis, especially those with few faults, the density can be very low. In spite of this, we believe that geochemical mapping using mineral and thermal waters can complement to a large extent chemical mapping using surface waters and rocks.

STUDY AREA

Portugal is a small country in South-West Europe. The mainland part of the country consists of the south-western one fifth of the Iberian Peninsula (Portuguese Mainland). Also part of Portugal are the two autonomous regions of Madeira and Azores, two archipelagos of islands in the Atlantic Ocean, which are not represented in the current study. The country is, especially when related to its surface area (ca. 92 000 km^2) and population (10 561 178, during the 2011 Census, (INE 2014)), particularly rich in mineral and thermal waters. Especially interesting is that many of these springs have been used for medicinal purposes (balneotherapy) since the earliest times, and still many ruins of Roman spas are present in the country (see *e.g.* ACCIAIUOLI 1952; 1953).

Geochemical mapping studies in Portugal are mainly restricted to two publications. Portugal took part in the Geochemical Atlas of Europe, which was carried out in 26 countries under the auspices of the Forum of European Geological Surveys (SALMINEN *et al.* 2005). A much higher density geochemical survey (at 1 site per 135 km^2) has been conducted by the University of Aveiro (INÁCIO *et al.* 2008). This study sampled A and O soil horizons from the Portuguese Mainland according to standards set by the IGCP (UNESCO's International Geoscience Programme) project 259 (DARNLEY & GARRETT 1990; DARNLEY *et al.* 1995). This resulted in the first Soil Geochemical Atlas of Portugal. Unfortunately, this publication presented maps with only little overlap with the present study.

4

In a recent study EGGENKAMP & MARQUES (2013) compared different methods to classify the mineral waters that are found in Portugal. The results of this study are presented in maps showing the distribution of the different types of mineral waters in Portugal (including the autonomous regions Madeira and Azores). The maps presented in this book, which are based on the same analytical data, show that the distribution of the individual chemical species in the mineral waters agree very well with the distribution of the different water types based on the combination of all the major elements in the waters.

From a geological point of view the Portuguese mainland is rather complex. Roughly it can be divided into two broad regions: the Western and Southern Meso-Cenozoic Basins (see *e.g.* RASMUSSEN *et al.* 1998; AZARÊDO *et al.* 2002; UPHOFF 2005) and the Iberian Massif (*e.g.* AIRES-BARROS 1979; 1989; CABRAL 1989; RIBEIRO *et al.* 1990; 2007).

In the Western Meso-Cenozoic Basins, thermal and mineral springs occur near faults originating from typhonic valleys in areas of gypso-saline diapirism. For geochemical mapping purposes the Lower Jurassic *"Margas da Dagorda"* formation is very important as it contains evaporites, mainly gypsum ($CaSO_4.2H_2O$) and halite ($NaCl$), which can be dissolved by the thermal and mineral waters that evolve to Na^+, Ca^{2+}, Cl^- and SO_4^{2-}-rich waters. Due to their location at the edge of the Atlantic spreading zone the sedimentary rocks in these regions contain many faults. In the Southern Meso-Cenozoic Basin, the thermal and mineral waters belong essentially to the $Na-HCO_3$ and $Ca-HCO_3$-type due to their interaction with igneous (*e.g.* nefeline syenites) and sedimentary (*e.g.* limestone) rocks respectively.

The central and northern parts of the Portuguese mainland (the Iberian Massif) consist mainly of Proterozoic and Paleozoic rocks (mainly granitic and schistose rocks), which were folded heavily during the Hercynian orogeny. Especially in the northern part of the country very large amounts of granitic rocks intruded the sedimentary-schistose rocks, causing metamorphism (DALMEYER & MARTINEZ GARCIA 1990). This region is also rich in quartziferous dikes rich in sulphide mineralisations. This region is rich in $Na-HCO_3$-rich sulphurous waters, weak in mineralisation, as well as in some $Na-HCO_3-CO_2$-rich waters with higher mineralisation. Hyposaline waters (weakly mineralised) are associated with geological environments where quartziferous rocks predominate (*e.g.* AIRES-BARROS 1979).

In the central part of the country there is a large basin, drained by the rivers Tagus and Sado. This basin consists of thick layers with young, mainly Neogene sediments with almost no faults. Due to its geological structure this area is very poor in mineral and thermal springs.

Map 1 displays a general overview of the geology of the Portuguese Mainland, showing clearly the presence of large sinistral faults, most of them with a NW-SE and NNE-SSW trend, along which most mineral and thermal waters issue (*e.g.* CABRAL 1989; RIBEIRO *et al.* 1990; 2007). The Map also shows the locations of the most important cities, such as the capitals of the districts.

As Portugal has a very high density in mineral and thermal waters it is a very good location to test the hypotheses that these water compositions can be used for geochemical mapping. The total number of thermal and mineral springs in Portugal is difficult to estimate. Its number depends partly on the definition used, but in general springs considered as "thermal" and "mineral" are those that are known for a long time. For example springs described by HENRIQUES (1728) and still known today are certainly part of the thermal and mineral water compilation. They should also have a "medical" (therapeutic) application, which is actually the case for most long standing thermal and mineral springs in Portugal, although we have to realise that in most cases for smaller springs this therapeutic use is based only on local tradition.

A modern inventory, partly based upon HENRIQUES (1728) has been published by BASTOS (2008) on the Internet. This inventory describes a total of 650 locations on the Portuguese mainland where natural water discharges that still has or had in the past any therapeutic use (either on a local, regional or national scale). This inventory was used as the basis for the definition of "thermal / mineral spring" on the Portuguese mainland used in this publication.

MATERIALS AND METHODS

Sample data

For this study no new chemical analyses of thermal and mineral waters were made. To be able to study the distribution of thermal and mineral waters of the Portuguese Mainland as many as possible data, which were all found in publicly available literature, were collected. Data for 491 locations were found in literature available from the Instituto Superior Técnico (IST) library and the Internet. The total number of springs from which geochemical analyses were found was 606, as in several locations, especially at important "Spas", various springs (and a limited number of boreholes) are discharging thermal or mineral water with, in general, comparable water compositions. Unfortunately, as can be seen in *Table 1*, both the springs and their analyses are not uniformly distributed over the country.

In *Table 1* it is shown that the density of thermal or mineral water locations varies between about 1 per 100 km^2 in the northern part of the country, and about 1 per 300 km^2 in the southern part of the country. The number of analyses of spring locations, in this *Table*, is compared to the number of spring locations as published by BASTOS (2008). There are some discrepancies here, as the number of reported analyses in the literature is also unevenly spread over the country. In any case from these data it is clear that the density of mineral spring locations is lower in the south than in the north. The actual locations of the springs can be estimated from *Maps 2 to 16* where they are visible as small dots on the maps. The distribution of these dots also indicate that the distribution of springs is uneven, showing more dots in the north than in the south of the country.

Table 1: *Distribution of thermal and mineral waters in the main regions of the Portuguese mainland. Northern region: districts of Viana do Castelo, Braga, Porto, Vila Real and Bragança; Central region: districts of Aveiro, Viseu, Guarda, Coimbra and Castelo Branco; Western region: districts of Leiria, Lisboa and Santarém and Southern region: districts of Portalegre, Évora, Setúbal, Beja and Faro. The districts are named after their capitals which are indicated in Map 1.*

Region	Area (km^2)	No. of springs	Density (#/100 km^2)	No. of analyses	Density (#/100 km^2)
Northern	18259	252	1.38	197	1.08
Central	23955	155	0.65	132	0.55
Western	13023	62	0.48	43	0.33
Southern	33707	137	0.41	119	0.35
Portuguese mainland	88944	606	0.68	491	0.55

The number of analysed samples in the north is so high mainly due to the comprehensive number of analyses published in the four volume series of the *"Inventario Hidrológico de Portugal"* (D´ALMEIDA & DE ALMEIDA 1966; 1970; 1975; 1988). It is very unfortunate that no more volumes in this series, covering the central and southern parts of the country, were prepared due to the dead of the authors. In this series, systematic sampling and analysis of all thermal and mineral sources was done. For the rest of the country, analyses have been taken from classic publications on therapeutic, mineral and thermal waters (LUZES *et al.* 1930; 1934; 1935; ACCUAIOLI 1952) and reports for individual springs (*e.g.* PIMENTEL 1852; LEPIERRE & HERCULANO DE CARVALHO 1930). Together, these publications cover the period from about 1850 until 1970. More recent data are taken from papers describing individual thermal and mineral water locations (*e.g.* AIRES-BARROS *et al.* 1998; CALADO & CHAMBEL 1999; FERREIRA GOMES *et al.* 2001;

MARQUES *et al.* 2001; 2003; 2006; 2008; 2010; LOPO MENDONÇA *et al.* 2004; CARREIRA *et al.* 2004, DA COSTA *et al.* 2006; MATIAS *et al.* 2009), and from a collection of new data compiled by the Geological Survey of Portugal to commemorate new legislation on mineral and thermal waters (DGGM 1992). Unfortunately, only few data in the Alentejo region (in the southern part of the country) were found in the open literature. So we had to revert to ERHSA (2000), a report describing springs in the Alentejo region which are mostly not the same as defined by BASTOS (2008) but defined as those springs and fountains (*"fontes"*) which are used for human consumption as drinking water and have an average of more than 50 visitors per day. The exact locations and the chemical analyses of the springs were not given in this report but graphically determined from graphs and maps available in it. Chemical analyses for all parameters, except for total dissolved solids (TDS), were used as published in the respective reports. The TDS has been calculated simply by adding the reported values in mg/L for all cations, anions and SiO_2, but excluding gaseous parameters such as CO_2, O_2 and N_2. The definition of carbonate species has been changing over time. Originally it was normally reported as CO_3^{2-}, not taking into account the fact that HCO_3^- is the most stable carbonate species in most water samples. In more recent analyses, either CO_3^{2-} or HCO_3^- was reported depending on the pH of the thermal or mineral waters. The analyses were corrected for this effect and the carbonate map was produced using the sum of the carbonate species (ΣCO_3 or $HCO_3^- + CO_3^{2-}$).

From some springs, especially those of the more well-known spas we were able to obtain more than one analysis, often conducted at different times. For example FERREIRA GOMES *et al.* (2001) showed several historic analyses from the main source at São Pedro do Sul (district of Viseu). Several other well-known springs have been reported as well as in classic works (LUZES *et al.* 1930, 1934, 1935; ACCIAIUOLI 1952) as in modern studies. The results of the analyses, sometimes conducted a century apart, are very much comparable. Actually, the only analytes where significant differences are found between older and younger analyses are the analyses of F^- and SO_4^{2-}. Apparently in older analyses it was not possible to properly analyse fluoride and a large part of it was interpreted as sulphate. The springs which are affected by this effect show relatively low fluoride data and high sulphate in older analyses while in younger analyses fluoride is significantly higher and sulphate lower. For example in the case of São Pedro do Sul fluoride is 2.2 mg/L and sulphate 25 mg/L in the old analysis reported in ACCIAIUOLI (1952), while they are about 17 mg/L (F^-) and 9 mg/L (SO_4^{2-}) in younger analyses (FERREIRA GOMES *et al.* 2001). This effect only affects samples with high fluoride contents. As no problems are found with any other analyte only for fluoride and sulphate the old analyses have been removed from the database.

Except for silica all studied species have concentrations which variations can vary several orders of magnitude from one region to another. *Table 2* shows some important descriptive statistics of the species from which contour maps are presented in this study. Most species show a large positive skewness. This indicates that most species are tailing in the high concentration range, which is shown in the median and percentile concentrations. For this reason it was decided to produce the maps using the logarithm of the mg/L values. As most chemical species show a concentration distribution spanning several orders of magnitude this is the only way to present the distribution of the data on the maps in a way that it is properly understood in all areas.

Table 2: Descriptive statistics for the chemical species and total dissolved solids used in this study. All values (except skewness) are reported in mg/L, and the pH is reported in pH values. St.Dev. means Standard Deviation, Skew. means Skewness. 10%, 25%, 75% and 90% are the respective percentiles of the distribution while Median is the value where 50% of the observations is below, and 50% above. Min. and Max. indicate respectively the lowest and highest observation in this study.

	Average	St. Dev.	Skew.	Min.	10% P.	25% P.	Median	75% P.	90% P.	Max.
pH	6.92	1.17	0.34	2.50	5.61	6.08	6.93	7.50	8.50	11.40
TDS	1370	10012	18.9	18.4	54.7	102	278	646	2434	220000
Na^+	379	3781	19.5	2.1	6.6	12.0	34.2	108	542	83900
Cl^-	513	5906	19.3	1.1	6.4	12.1	24.9	57.9	203	130000
Ca^{2+}	54.9	140.5	6.6	0.2	1.2	3.4	11.4	52.7	133	1590
Mg^{2+}	16.5	38.8	6.6	0.02	0.4	1.2	3.7	16.9	39.1	461
SO_4^{2-}	80.1	280.3	6.1	0.06	2.1	5.8	12.8	36.1	95.7	3080
ΣCO_3	264.8	600.1	4.9	0.7	9.5	22.0	94.5	246	418	5420
SiO_2	27.6	21.4	1.3	0.5	7.1	12.0	20.7	37.6	61.7	140
NO_3^-	12.8	37.5	6.4	0.01	0.1	0.1	1.6	8.0	25.0	498
Al^{3+}	0.70	9.34	20.3	0.01	0.01	0.01	0.01	0.05	0.32	21
Fe^{2+}	2.2	28.7	23.7	0.01	0.02	0.04	0.08	0.4	2.8	699
HS^-	1.2	6.5	14.0	0.01	0.01	0.01	0.01	0.06	2.8	127
F^-	2.7	5.6	2.3	0.01	0.05	0.05	0.05	1.7	12.0	28
Li^+	0.34	1.26	11.3	0.01	0.01	0.01	0.01	0.11	0.97	23

When evaluating this Table, it is important to realise that pH already is a unit that is characterised by a logarithm. This is the reason that for the pH the standard deviation and the skewness are so much smaller than for the other species

Preparation of the geochemical maps

Geochemical maps have been produced using Golden Software's program SURFER Version 5.02. Contour maps were prepared based on the logarithm of the mg/L values obtained from our literature sources. Maps were made using the default Kriging gridding method, which is the recommended gridding method for datasets which are not too large and it produced very satisfactory maps as is shown in *Maps 2* to *16*. In these maps the contours were smoothed. Initially maps for parameters with only few missing values ($< 4\%$ of total analyses; pH, SiO_2, Na^+, Mg^{2+}, Ca^{2+}, Cl^-, SO_4^{2-} and ΣCO_3) were prepared. As the distribution of other species is also of much interest, attempts were made to produce maps for species from which it could be sure that the missing values in the dataset were due to the fact that they were below their detection limits, for example because it was mentioned in the report, or by comparison of related samples. For $Fe(T)$, NO_3^-, F^-, Li^+, Al^{3+} and HS^- (between 11 and 78% missing values) the missing values were replaced with a fixed, constant low value at the level of the lowest measured concentration, which was for most species between 0.005 and 0.01 mg/L as 0.01 mg/L in most cases is the lowest concentration that is reported for these species. Maps were then produced using the combination of the collected measured data and assumed "lower than detection limit" data. Acceptable maps could be produced which at least make it possible to asses where these rarer species are found in significant concentrations. When more data will be available in the future there is a risk that these maps change, especially in those regions rich in samples with concentrations lower than the detection limit. We are confident however that the regions with higher concentrations are mapped properly.

It was not possible to produce a map with the K^+ distribution. This is due to the lack of analyses in older publications, and especially as D´ALMEIDA & DE ALMEIDA (1966; 1970; 1975; 1988), the most important source for the smaller and lesser important thermal and mineral springs, did not analyse K^+ as one of their standard parameters, and K^+ is missing in roughly half of the springs. Unlike the species discussed above it is not possible to replace missing K^+ values with lower than detection limit values, as many of these springs must have significant K^+ concentrations. Production of a K^+ map will only be possible after enough new data are collected in the future, especially in the northern part of the country.

HYDROGEOCHEMICAL SIGNATURES OF THE PORTUGUESE MAINLAND

On the Portuguese Mainland, mineral waters with temperatures above 20 °C are considered thermal waters and could potentially be used as geothermal resources (LOURENÇO & CRUZ 2005). However, most mineral waters on the Portuguese mainland are cold, with temperatures below 20 °C. Slightly warmer waters are found in the southern and western parts of the country, while warm waters, with temperatures above 40 °C emerge in a NW-SE stretching region east of Porto city, with thermal waters showing also a very interesting mineralization, such as São Pedro do Sul (FERREIRA GOMES et al. 2001). In general, there is no direct relationship between the temperature of a spring and its chemical composition. This is very nicely shown in the Na-HCO_3-CO_2-rich type of waters found along the NNE-SSW-trending Verín-Régua-Penacova fault in Northern Portugal (e.g. MARQUES et al. 1998; 1999a; 2006), where several thermal and mineral waters with very similar geochemical signatures are found. Most of these mineral waters have low temperatures of about 17 °C, but among them also the hottest thermal water from the Portuguese mainland (i.e. Caldas de Chaves, district of Vila Real, with a temperature of 76 °C) is found. In this type of water, low temperatures enhance water-rock interaction since the solubility of deep seated CO_2 of mantle origin (MARQUES et al. 2001; CARREIRA et al. 2010) in water increases with decreasing temperature (GREBER 1994). The mantle origin of CO_2 in these waters is shown through the carbon isotope data (about -6 ‰) of the CO_2, $^3He/^4He$ ratios (R/Ra is between 0.5 and 2.68) and the $CO_2/^3He$ ratios (~10^9; CARREIRA et al. 2010). The fact that in this region water-rock interaction is mainly governed by CO_2 rather than by high temperatures explains why some of the most mineralised waters in Portugal are cold waters.

Distribution of chemical elements in thermal and mineral waters

Total Dissolved Solids and pH

The main variation in distribution of the thermal and mineral waters can be shown in a map of the total dissolved solids (TDS, Map 2). Large differences are shown over the country, covering variations spanning about five orders of magnitude. The TDS content of the thermal and mineral waters is dependent on four main factors: i) the chemical composition of the rocks, ii) the water-rock interaction temperature, iii) the intensity of water-rock

TDS (log of mg/L)

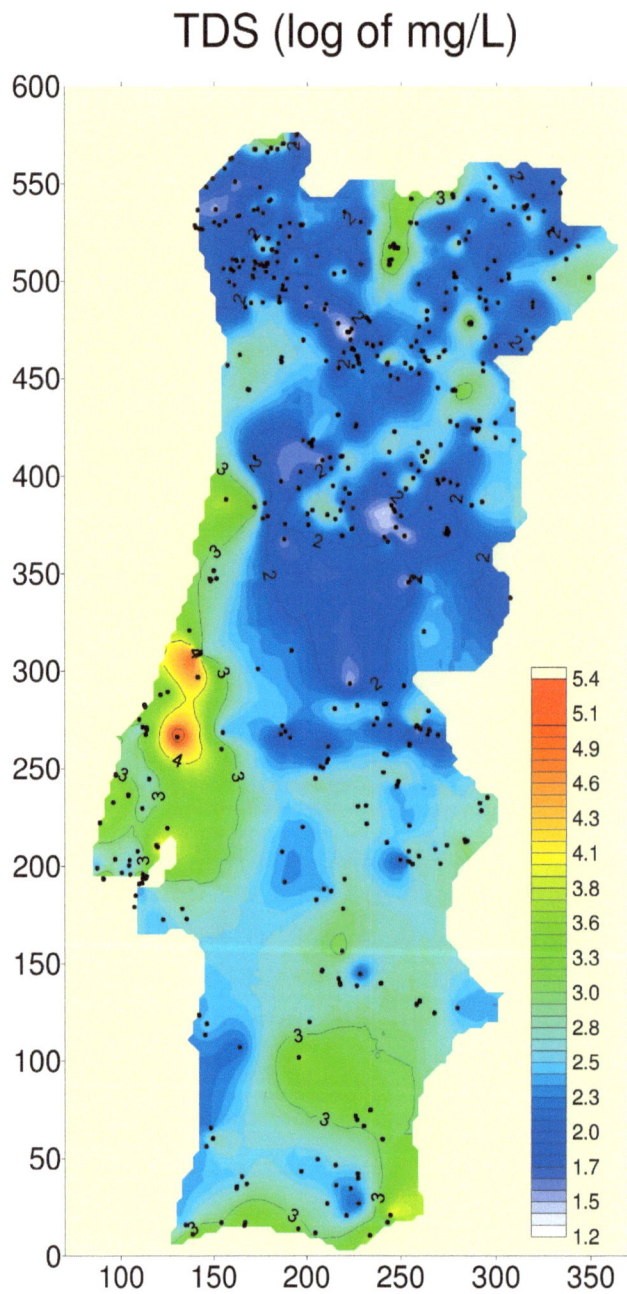

Map 2: Total dissolved solids (TDS) distribution in Portuguese thermal and mineral waters. The numbers on the lines in the Map indicate the logarithm of the concentrations in mg/L (i.e. 1 means 10 mg/L; 2 means 100 mg/L). The dots reflect the locations of the springs from which analyses have been used in the calculations of the maps. Scales on the x- and y-axis are in km. The maximum value of 5.4 equals to 250000 mg/L, the minimum value of 1.2 to 16 mg/L.

interaction between the water and the rocks and iv) the presence of deep seated (upper mantle) CO_2. These effects are also evident on the geochemical maps of individual chemical species. It is the combination of the chemistry of the rocks with tectonic effects occurring at faults that show the usefulness of thermal and mineral water geochemical maps.

Photo 1: The Queen Leonor Thermal Hospital in Caldas da Rainha (district of Leiria). The thermal water of Caldas da Rainha is a Na-Cl-HS water with a high TDS value.

The highest TDS values are found in the Western Meso-Cenozoic Basin, in areas where salt deposits from the Lower Jurassic "Margas da Dagorda" formation do occur. At some locations in this basin salt diapirs do approach the surface very closely and these are the locations where the mineral waters with the highest dissolved load are found, such as for example in Rio Maior, district of Santarem, where the diapir is so shallow that this spring ("Fonte da Pipa") contains 137 g/L TDS of which 98% is NaCl. As a result of the presence of relative shallow halite deposits circulating groundwaters are less prone to dilution by shallower groundwaters on their way to the surface explaining why especially in these regions mineral water with very high NaCl concentrations are found. This high NaCl concentration makes it possible that this spring is used to extract and commercialise salt (SCHMIDT 2001; EGGENKAMP et al. 2010; 2013).

The highest value in the Algarve region is found in the eastern part of the Southern Meso-Cenozoic Basin in the "Fonte Salgada" with a TDS of over 12 g/L. Other areas with relatively high TDS values are found in the southern Alentejo region, in the Iberian Massif. A possible explanation for these relatively high TDS values might be the combination of relatively reactive magnesium and carbonate rich rocks with the high air temperatures in the summer months and relatively shallow water circulation. These conditions

13

pH

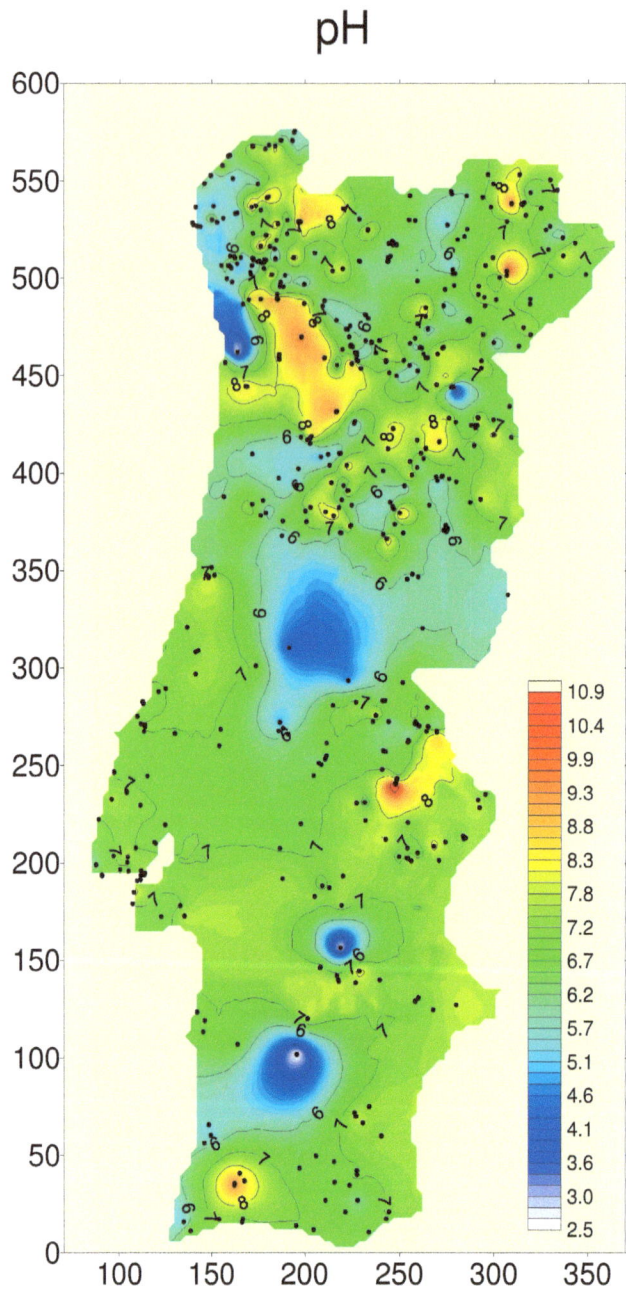

Map 3: Distribution of pH in thermal and mineral waters from the Portuguese mainland. In this Map the numbers on the lines reflect actual pH values. The dots reflect the locations of the springs from which analyses have been used in the calculations of the maps. Scales on the x- and y-axis are in km.

14

might be responsible for evaporation of water, which also dissolves the rock more easily than in the north leading to a higher TDS in the waters. On the contrary low TDS values are mostly found in the northern half part of the country. This trend is due i) to climate features (a very wet region), and ii) to the occurrence of granitic rocks in huge parts of this region, leading to relatively low water-rock interaction, but often very interesting HS, F and Li-rich waters. The exceptions to this low mineralization are a few relatively small areas where thermal and mineral waters related to very deep faults obtain their CO_2 from the upper mantle (MARQUES *et al.* 2001; 2006; CARREIRA *et al.* 2010) which results in intensive water-rock interaction with the local granites and produce the very important $Na-HCO_3-CO_2$-rich waters which are found in the Chaves-Pedras Salgadas, Sandim and Vila Flor areas (districts of Vila Real, Porto and Bragança, respectively) (D'ALMEIDA & DE ALMEIDA 1970; AIRES-BARROS *et al.* 1995; 1998; MARQUES *et al.* 2001; 2006).

Photo 2: The spring "Ermida" in Cabeço de Vide (district of Portalegre), the mineral spring with the highest pH encountered in Portugal.

The pH of the thermal and mineral waters (*Map 3*) also is an important characteristic. The pH defines several of important chemical reactions that can take place between the waters and the rocks. For example at low pH values many, especially transition, metals are much more soluble than at neutral pH values. Most mineral and thermal waters in the Portuguese mainland have neutral pH values (6 < pH <8, see also *Table 2*). Very few locations with extreme pH values are found, and these are characterised on the map (*Map 3*) as either light areas for very acid springs, such as Aljustrel, district of Beja, (PIMENTEL 1852) or dark areas for very alkaline springs such as Cabeço de Vide, district of Portalegre, (MARQUES *et al.* 2008). Only in the north-western part of the country a larger area with alkaline springs are found: these are the HS rich mineral waters, of which São Pedro do Sul, district of Viseu, is an important example and that have pH values close to 9.

Based upon the spatial distribution of the chemical species in the studied area they can be grouped depending on their geochemical characteristics. The differences between these maps can be described and explained with our knowledge of the geology, geochemistry and hydrogeology of the country.

Sodium and chloride

Comparing the two maps representing the distribution of Na^+ and Cl^- (*Maps 4* and *5*, respectively) in the thermal and mineral waters it can be observed that, while in the southern and western parts of the country the two maps are more or less similar; in the northern part of the country large regions with very low Cl^- concentrations are found, while Na^+ is significantly more concentrated.

Photo 3: The "buvette" inside the Queen Leonor Thermal Hospital in Caldas da Rainha.

These differences can be explained by the large variations in the geological scenery and tectonic structures of the two parts of the country. As mentioned above, in the Southern and Western Meso-Cenozoic Basins both Na^+ and Cl^- are controlled by dissolution of Lower Jurassic salt deposits. On the contrary, in the northern part of the country granite rocks dominate the geological environment. As granitic rocks contain more Na^+ than Cl^-, in this area water-rock interaction results in relatively increased Na^+ contents in waters due to hydrolysis of plagioclases. In most thermal and mineral waters CO_2 balances the Na^+ and HCO_3^-, resulting in $Na-HCO_3$ dominated waters.

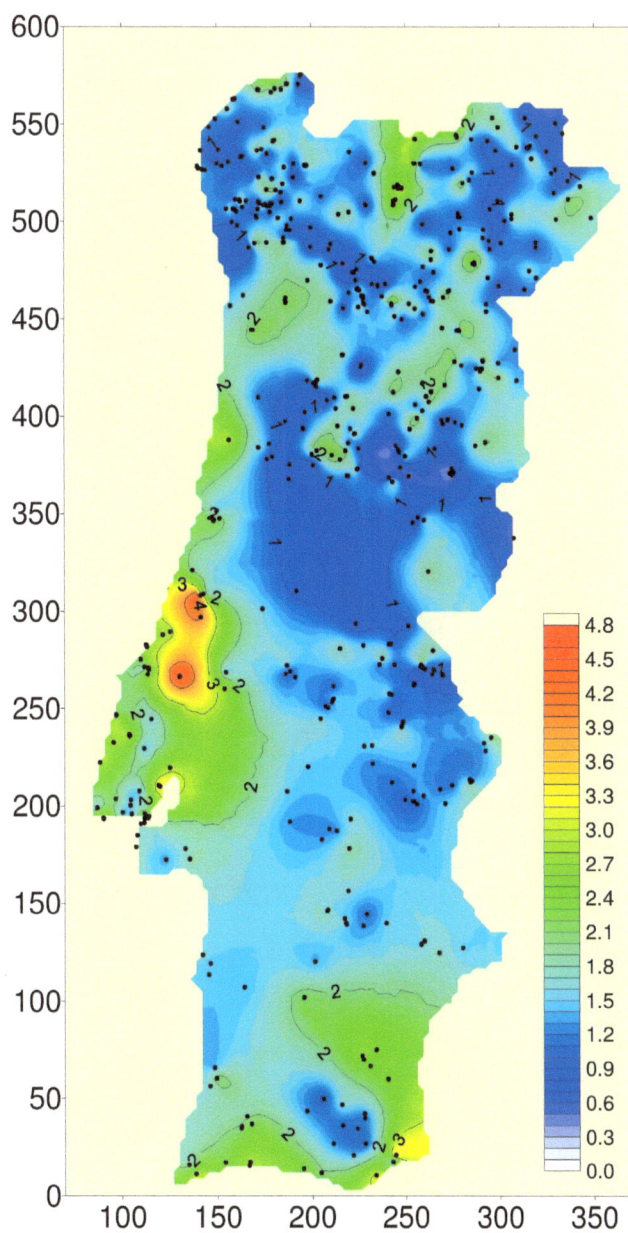

Sodium (log of mg/L)

Map 4: Distribution of Na⁺ in thermal and mineral waters from the Portuguese mainland. The numbers on the lines in the Map indicate the logarithm of the concentrations in mg/L (i.e. 1 means 10 mg/L; 2 means 100 mg/L). The dots reflect the locations of the springs from which analyses have been used in the calculations of the maps. Scales on the x- and y-axis are in km. The maximum value of 4.8 equals to 63000 mg/L, the minimum value of 0.0 to 1 mg/L.

Chloride (log of mg/L)

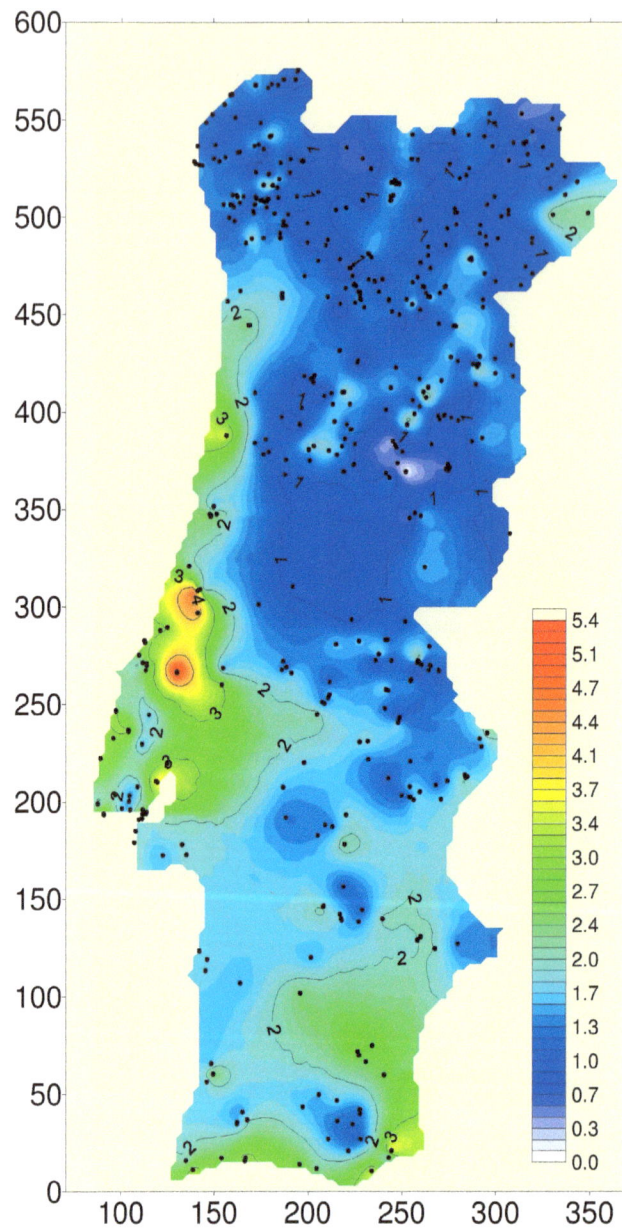

Map 5: Distribution of Cl⁻ in thermal and mineral waters from the Portuguese mainland. The numbers on the lines in the Map indicate the logarithm of the concentrations in mg/L (i.e. 1 means 10 mg/L; 2 means 100 mg/L). The dots reflect the locations of the springs from which analyses have been used in the calculations of the maps. Scales on the x- and y-axis are in km. The maximum value of 5.4 equals to 250000 mg/L, the minimum value 0.0 to 1 mg/L.

18

Along the deepest faults, which are mainly NNE-SSW trending, the amount of CO_2 escaping from the upper mantle (MARQUES *et al.* 2001; 2006; CARREIRA *et al.* 2010) can be so high that the waters are supersaturated in CO_2 and can be described as Na-HCO_3-CO_2-rich type waters (*e.g.* MARQUES *et al.* 2006). As Na^+ is the main cation in the majority of thermal and mineral waters found on the Portuguese mainland normally a good correlation between the TDS and the Na^+ maps for the Portuguese mainland is observed (see *Maps 2* and *4*, respectively). Only in the northern part of the Alentejo region (southern half of the Iberian Massif) Ca^{2+} is the dominant cation in most thermal and mineral waters, resulting in one of the few discrepancies between the TDS and Na^+ maps.

Calcium, magnesium, sulphate and (bi)carbonate

Ca^{2+} and Mg^{2+} are related cations that show a very common behaviour in the Portuguese mainland thermal and mineral waters. As shown in *Maps 6* and *7* their distribution is very similar. Roughly, three regions can be recognised: high concentrations in the Western Meso-Cenozoic Basin, low concentrations in the northern part of the Iberian Massif and intermediate concentrations in the southern part of the Iberian Massif and in the Southern Meso-Cenozoic Basin. The high concentrations in the Western Meso-Cenozoic Basin are the result from dissolution of evaporite minerals from the Lower Jurassic *Margas da Dagorda* formation. In the southern part of the Iberian Massif Ca^{2+} and Mg^{2+} concentrations are intermediate due to dissolution of carbonate minerals, mainly from metamorphosed carbonate rocks (marls). In this region the Mg^{2+} concentrations are, compared to Ca^{2+}, relatively higher than in the western part of the country. In this region the sedimentary and metamorphic carbonate rocks contain a relatively larger amount of Mg^{2+} related to Ca^{2+} than in the western part of the country, where Ca^{2+} concentrations are higher due to the presence of gypsum. It should also be stated that in the southern part of the Iberian Massif, at the Ossa Morena Zone of the Iberian Hercynian Belt, the Lower Cambrian carbonate sequence was intruded and metamorphosed by mafic and ultramafic rocks, forming a NW-SE cumulate-type structure of Ordovician age, which has been subjected to serpentinization - rodingitization processes (COSTA *et al.* 1993), contributing to the presence of high Ca^{2+} concentrations in some local mineral waters (MARQUES *et al.* 2008). In the northern part of the Iberian Massif both Ca^{2+} and Mg^{2+} show very low concentrations (<5 mg/L). This is the result of water-rock interaction between water and calcium/magnesium-poor rocks (such as granites and schists). In the north-easternmost part of the country ultramafic Ca-rich rocks are present, which is reflected in the slightly higher Ca^{2+} and Mg^{2+} concentrations in thermal and mineral waters from that area (RIBEIRO *et al.* 2007).

19

Calcium (log of mg/L)

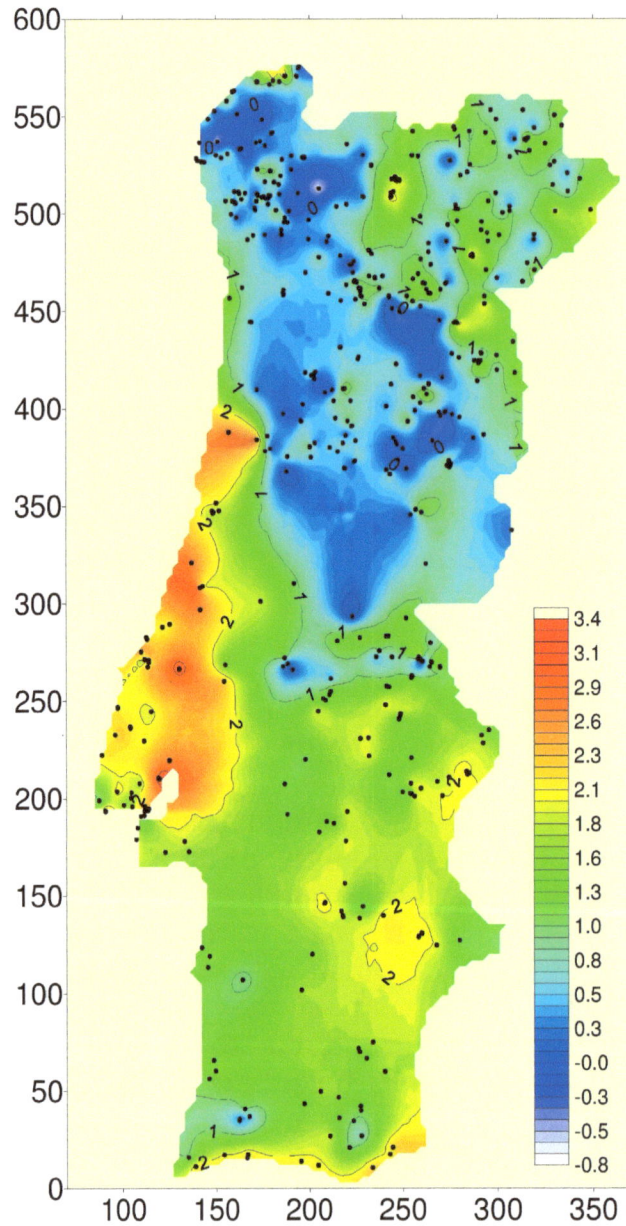

Map 6: *Distribution of Ca²⁺ in thermal and mineral waters from the Portuguese mainland. The numbers on the lines in the Map indicate the logarithm of the concentrations in mg/L (i.e. 1 means 10 mg/L; 2 means 100 mg/L). The dots reflect the locations of the springs from which analyses have been used in the calculations of the maps. Scales on the x- and y-axis are in km. The maximum value of 3.4 equals to 2500 mg/L, the minimum value of -0.8 to 0.16 mg/L*

20

Magnesium (log of mg/L)

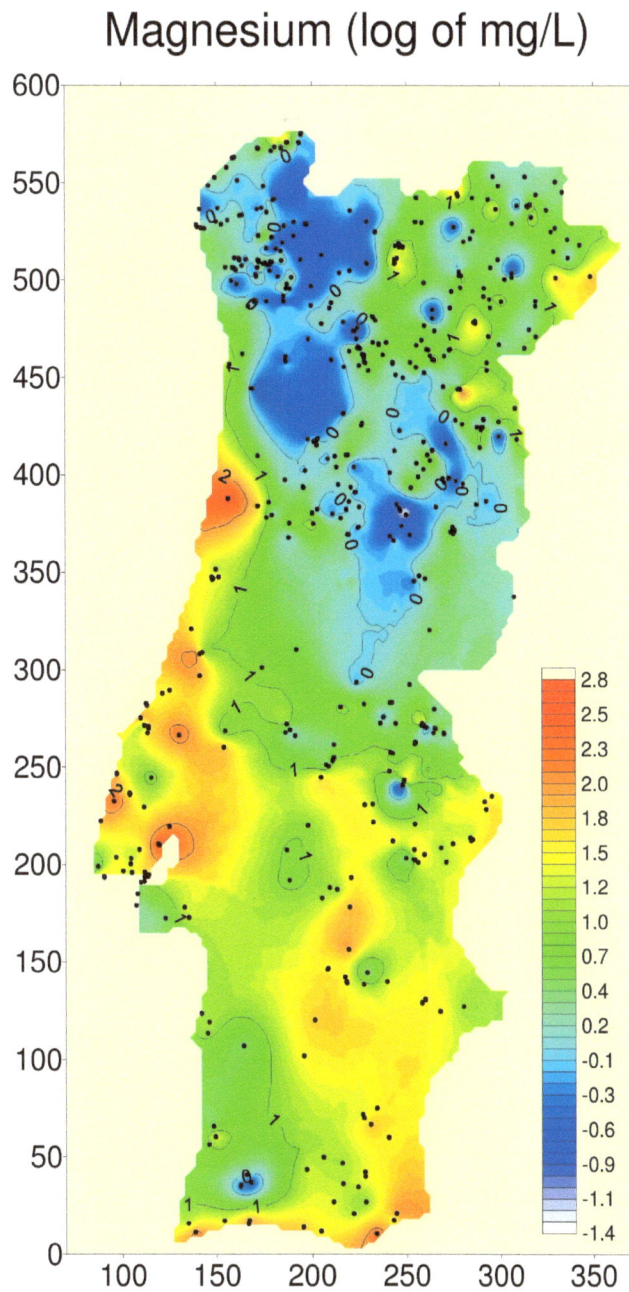

Map 7: Distribution of Mg^{2+} in thermal and mineral waters from the Portuguese mainland. The numbers on the lines in the Map indicate the logarithm of the concentrations in mg/L (i.e. 1 means 10 mg/L; 2 means 100 mg/L). The dots reflect the locations of the springs from which analyses have been used in the calculations of the maps. Scales on the x- and y-axis are in km. The maximum value of 2.8 equals to 630 mg/L, the minimum value of -1.4 to 0.04 mg/L.

21

Sulphate (log of mg/L)

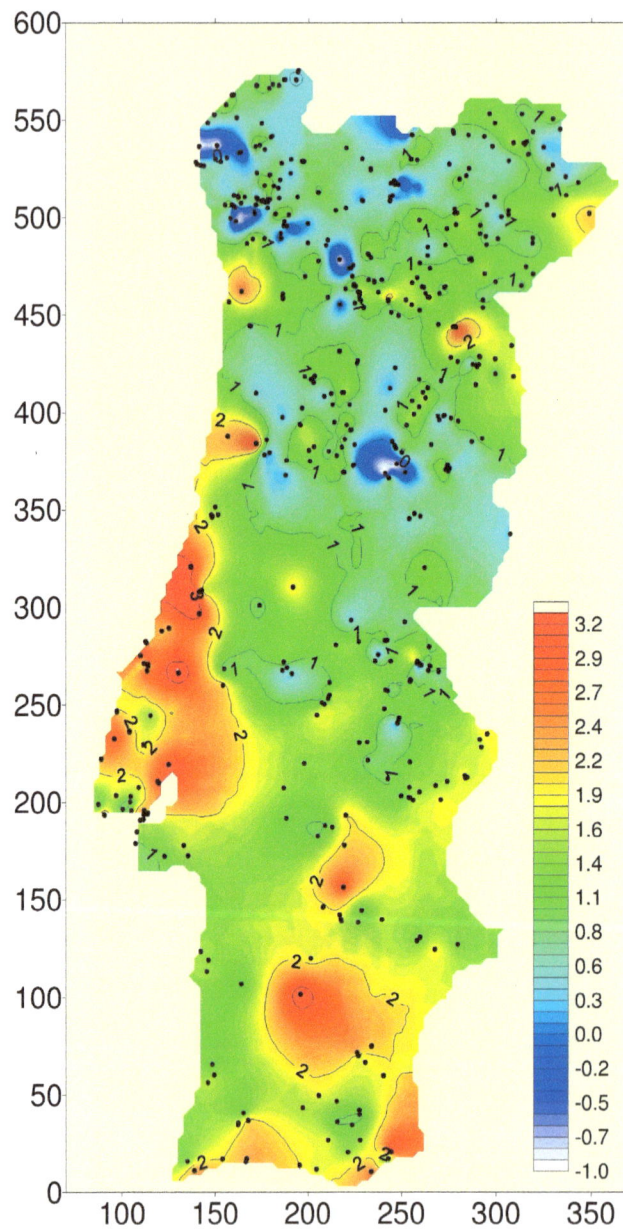

Map 8: Distribution of SO_4^{2-} in thermal and mineral waters from the Portuguese mainland. The numbers on the lines in the Map indicate the logarithm of the concentrations in mg/L (i.e. 1 means 10 mg/L; 2 means 100 mg/L). The dots reflect the locations of the springs from which analyses have been used in the calculations of the maps. Scales on the x- and y-axis are in km. The maximum value of 3.2 equals to 1600 mg/L, the minimum value of -1.0 to 0.1 mg/L.

The most soluble Ca and Mg-rich rocks are evaporites, which are mostly present in the Western Meso-Cenozoic Basin. The presence of these evaporite rocks explains both the high concentrations for the Ca^{2+} and Mg^{2+} cations as well as the SO_4^{2-} anion (*Map 8*) in this region.

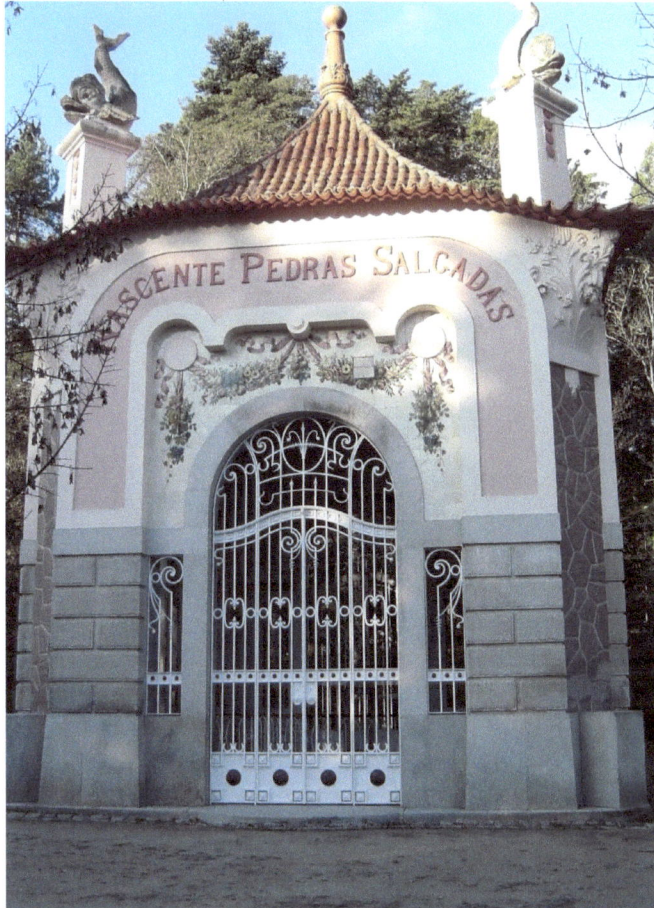

Photo 4: One of the springs of Pedras Salgadas (district of Vila Real), one of the most famous Na-HCO_3-CO_2 rich mineral waters in Portugal.

Except for areas were SO_4^{2-} is ascribed to evaporite rocks at depth, high SO_4^{2-} concentrations are also found in acid springs due to oxidation of sulphide minerals such as pyrite, as will be discussed in the section on the distribution of iron and aluminium. This could explain the high sulphate concentrations found in thermal and mineral waters from the southern part of the Iberian Massif, related to the so-called Iberian Pyrite Belt (*e.g.* SCHERMERHORN 1982; THADEU 1989; BARRIGA & CARVALHO 1983; ESPAÑA *et al.* 2005).

Ca and Mg carbonates have relatively comparable solubilities ($Ks_{CaCO3} = 10^{-8.42}$ and $Ks_{MgCO3} = 10^{-7.46}$). This explains that in regions with predominately carbonate dissolution such as the Alentejo region (southern part of the Iberian Massif) a relationship between Ca^{2+} and Mg^{2+} and carbonate species

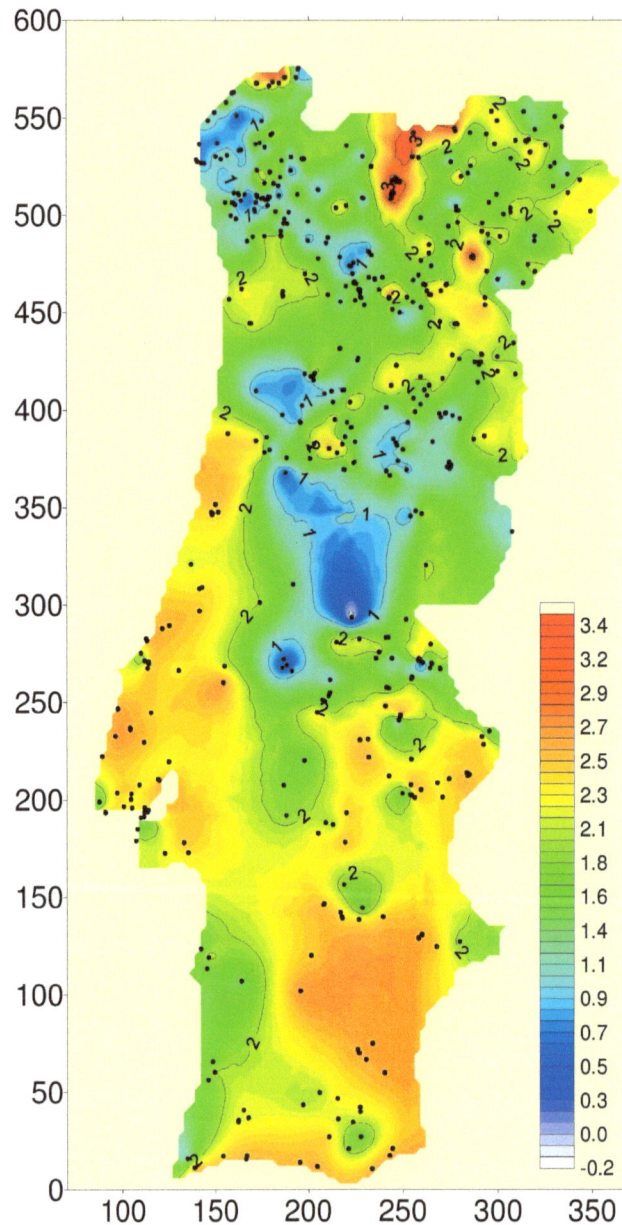

(Bi)carbonate (log of mg/L)

Map 9: Distribution of $HCO_3^- + CO_3^{2-}$ in thermal and mineral waters from the Portuguese mainland. The numbers on the lines in the Map indicate the logarithm of the concentrations in mg/L (i.e. 1 means 10 mg/L; 2 means 100 mg/L). The dots reflect the locations of the springs from which analyses have been used in the calculations of the maps. Scales on the x- and y-axis are in km. The maximum value of 3.4 equals to 2500 mg/L, the minimum value of -0.2 to 0.63 mg/L.

is found (*Maps 6, 7* and *9*). In this region, the highest Ca^{2+} and Mg^{2+} concentrations must be related to cold mineral waters, since the dissolution of Ca and Mg carbonates increases with decreasing temperatures (GREBER 1994), However, concentrations are still not very high in absolute terms as both Ca^{2+} and Mg^{2+} concentrations are in general between 10 and 30 mg/L.

The highest carbonate species concentrations (up to 3000 mg/L) are found in a small region in the northern part of the Iberian Massif (*Map 9*), where several Na-HCO_3-CO_2-type thermal and mineral waters can be found. These CO_2-rich thermal and mineral waters represent locally meteoric waters recharged at different altitudes (MARQUES *et al.* 1998; 2001; 2006; CARREIRA *et al.* 2010). Isotopic signatures of carbon present in these mineral water systems indicate a deep-seated (upper-mantle) origin for the CO_2. The most probable explanation by which the carbon dioxide could be transported from its deep source to the surface involves migration as a separate gas phase, through the extremely deep trending faults in these regions, being incorporated in the infiltrated meteoric waters (at considerable depth in the case of the hot – thermal - CO_2-rich waters and at shallow levels in the case of cold – mineral - CO_2-rich waters). Other solutes (such as HCO_3^- and Na^+) are originated from the local granitic rocks, being the dissolution favoured by the high CO_2 content of the circulating waters (MARQUES *et al.* 1998; 2001; 2006; CARREIRA *et al.* 2010). The CO_2 flux in this area can be as high as 110 $g/m^2/d$ (ANTUNES DA SILVA 2007), resulting in CO_2 concentrations that can be so high that the mineral waters are sometimes supersaturated with respect to CO_2 gas.

Silica

The variation in dissolved silica concentrations in the thermal and mineral waters on the Portuguese mainland is the smallest from all the species (*Map 10*, see also *Table 2*). The solubility of silica is mainly determined by the water-rock interaction temperature, which for example leads to high concentrations in volcanic spring wate rs in the Furnas area on the Azorean island of São Miguel (CRUZ *et al.* 1999).

Indeed, also on the Portuguese mainland the hottest springs are located in regions with relatively high silica concentrations. However, relatively high silica concentrations are also found in regions with only cold waters. One explanation could be that this water was originally warmer, and cooled on its way to the surface, or as the result of mixing with cold waters. Comparing SiO_2 data with the chalcedony geothermometer from ARNÓRSSON *et al.* (1983) revealed that this might be the case, as the warmest waters seem to have a temperature that is about 22 °C lower than the calculated chalcedony reservoir temperature. However, this also implies that the majority of other water samples have a larger deviation from the calculated temperature. In

Silica (log of mg/L)

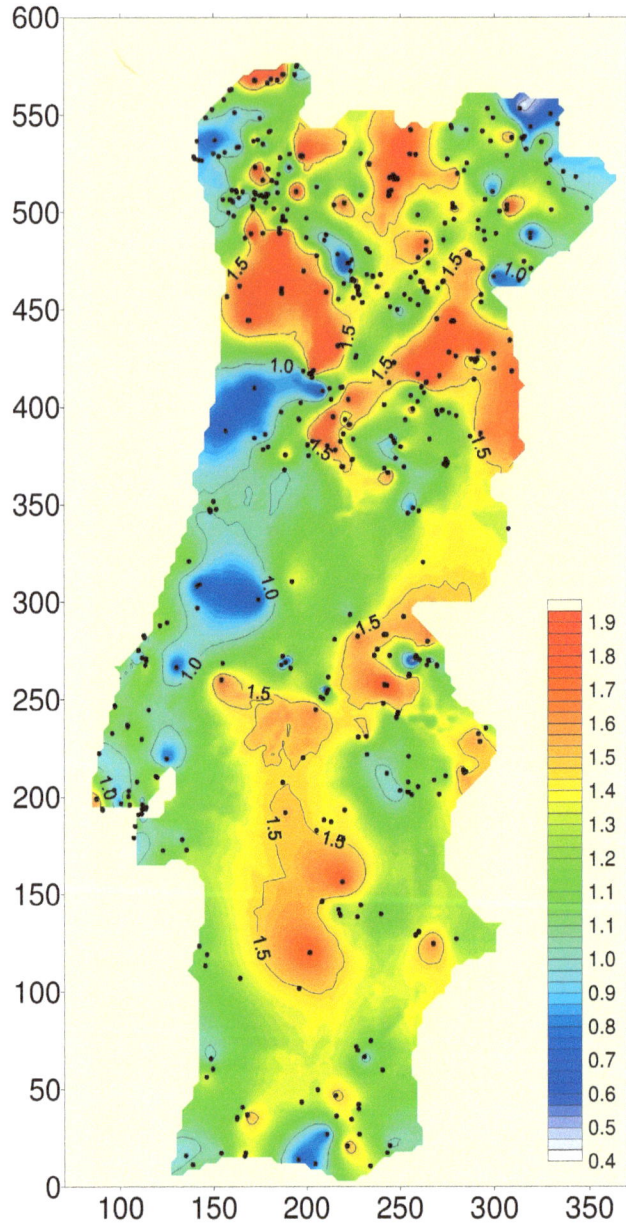

Map 10: Distribution of SiO₂ in thermal and mineral waters from the Portuguese mainland. The numbers on the lines in the Map indicate the logarithm of the concentrations in mg/L (i.e. 1 means 10 mg/L). The dots reflect the locations of the springs from which analyses have been used in the calculations of the maps. Scales on the x- and y-axis are in km. The maximum value of 1.9 equals to 79 mg/L, the minimum value of 0.4 to 2.5 mg/L.

26

general however water samples with higher temperatures have higher SiO_2 contents, although the SiO_2 concentration is higher than can be calculated from the temperature at the spring or borehole head. This is also confirmed in *Map 10* showing that most high SiO_2 regions are related with locations of high geothermal potential (*e.g.* AIRES-BARROS & MARQUES 2000; CABEÇAS 2010).

Photo 5: Melgaço (district of Viana do Castelo), the famous Ca-HCO₃-CO₂-rich mineral water from Northern Portugal also has one of the highest SiO₂ concentrations in Portugal.

Low SiO_2 concentrations are found in the Western Basin. Mineralization of mineral waters in this area is caused predominantly by dissolution of salt deposits of the *"Margas da Dagorda"* formation, which contain little silica, resulting in relatively low SiO_2 concentrations in mineral waters.

Nitrate

The map representing NO_3^- concentrations (*Map 11*) is probably the map with the slightest correlations with the geology of the country. In most samples nitrate is representative for anthropogenic sources mainly due to contamination related to the intensive fertilization of agricultural land. As nitrate is not retained in the mineral or organic soil phases it is easily transported with groundwater and, especially in cases where the deeper groundwater is not protected by impermeable layers, can contaminate deeper waters that discharge as mineral springs. Considering the distribution over the Portuguese mainland, it appears that in the north groundwater contamination is relatively rare and restricted to individual

27

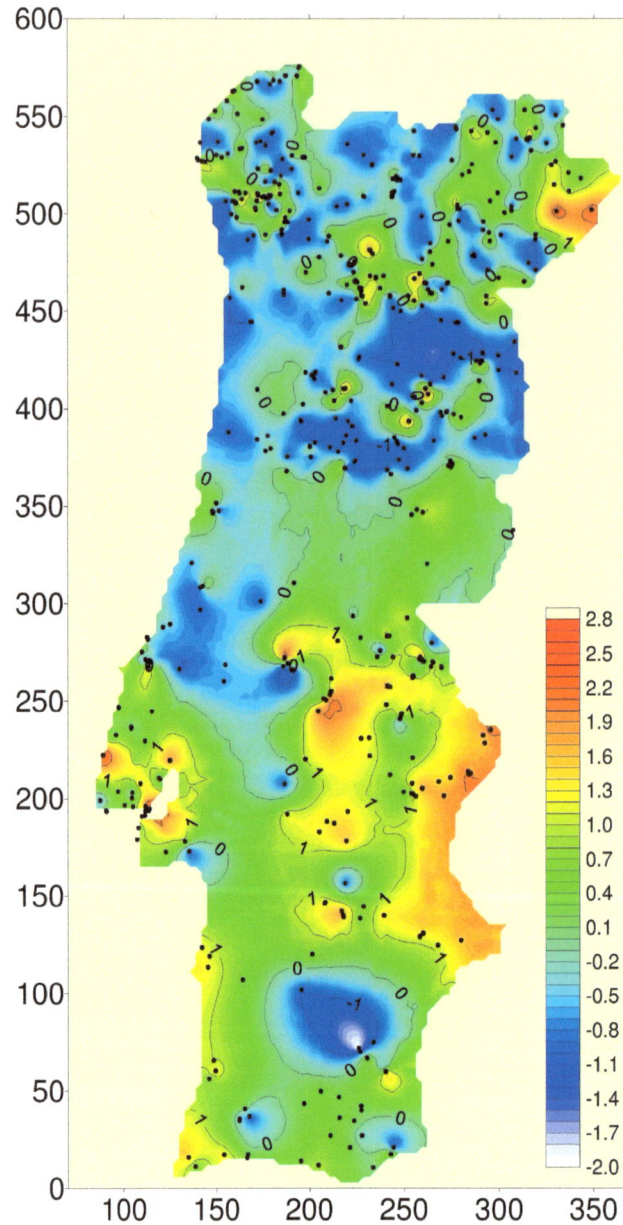

Nitrate (log of mg/L)

Map 11: Distribution of NO$_3^-$ in thermal and mineral waters from the Portuguese mainland. The numbers on the lines in the Map indicate the logarithm of the concentrations in mg/L (i.e. 0 means 1 mg/L; 1 means 10 mg/L). The dots reflect the locations of the springs from which analyses have been used in the calculations of the maps. Scales on the x- and y-axis are in km. The maximum value of 2.8 equals to 630 mg/L, the minimum value of -2.0 to 0.01 mg/L.

locations, while in the southern Alentejo region it is more common in the whole area. This may also indicate that thermal and mineral waters in this area follow a shallower and shorter circulation path. In the Lisbon area the nitrate content is also the result of urban groundwater contamination, which caused closure of most springs in the city in recent times (RAMALHO & LOURENÇO 2005). However, a small group of springs (the most well known is Termas de Santa Marta in Mafra municipality (district of Lisbon); LEPIERRE 1921) do have very high natural nitrate compositions. ACCIAIUOLI (1953) explained these high nitrate concentrations by possible bacterial oxidation of (reduced) nitrogen compounds, but as far as we know never any real research has been done into this issue.

Aluminium and iron

Al^{3+} and Fe^{2+} are related elements as they are much more soluble in acid environments than in neutral environments. In more alkaline environments (pH > 7) their solubility also increases again. For example in very alkaline environments high levels of aluminium as anionic complexes (AlO_2^-) can be present.

This is an especially important feature for Al as this is only found in the +III redox state. Iron exists in both the insoluble +III and the more soluble +II forms. In thermal and mineral waters iron is mostly found as Fe^{2+}, that after discharge of the water oxidises to insoluble Fe^{3+} which precipitates as a red deposit around the discharge point of the spring.

Map 12 shows the distribution of soluble Al species on the Portuguese mainland. The Al-rich springs can be recognised by the darker locations that do not have any regional distribution. These springs are also the springs with the lowest pH values. These locations are generally close to important mineral ore deposits and can be considered as natural acid mine drainage sites. Apart from this, the Al concentration is relatively high in the Western Meso-Cenozoic Basin where thermal and mineral waters are characterised by salt dissolution, and which have often high sulphate concentrations. In the northern part of the country the Chaves-Pedras Salgadas region is characterised by the presence of Na-HCO_3-CO_2-rich type thermal and mineral waters. In these areas Al^{3+} may form complexes with F^-. Al^{3+} forms strong complexes with F^- resulting in dissolved ions such as AlF^{2+} and AlF_2^+ which increase the solubility of Al also in near neutral solutions (HEM 1968). It is also found in soils where sorption of fluoride onto soils releases aluminium in solution (HARRINGTON *et al.* 2003). The presence of fluoride in water thus increases the solubility of aluminium at near neutral pH, which might explain the relatively high aluminium concentration in these areas. The presence of high sulphate concentrations (> 500 mg/L) will also

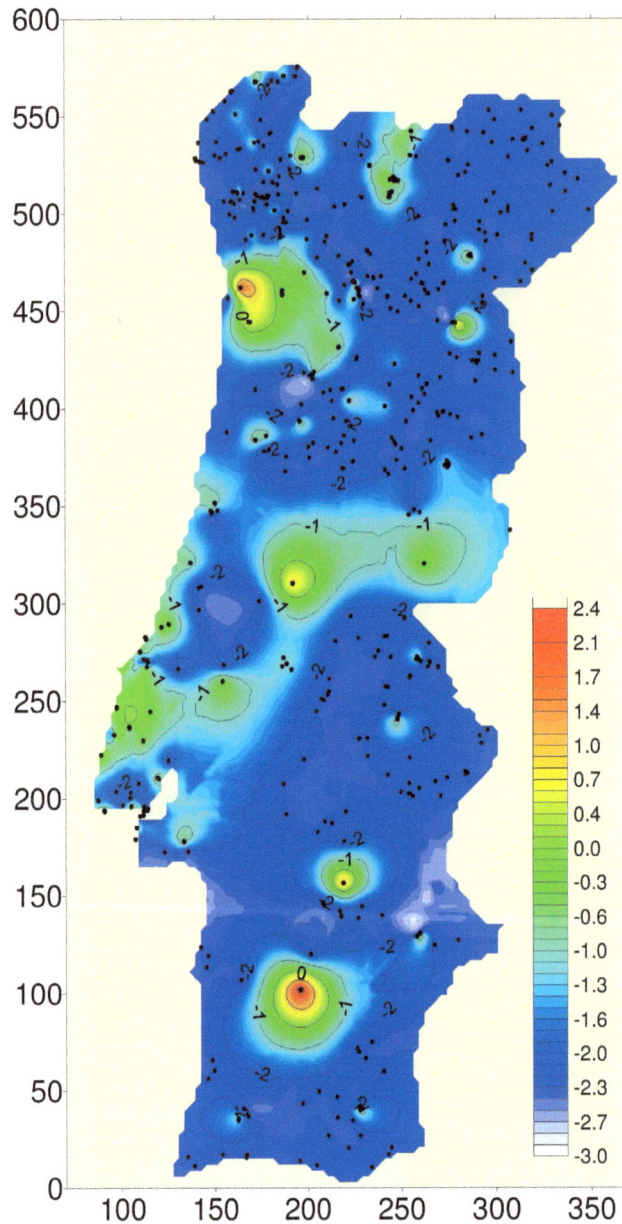

Map 12: Distribution of soluble Al species in thermal and mineral waters from the Portuguese mainland. The numbers on the lines in the Map indicate the logarithm of the concentrations in mg/L (i.e. 0 means 1 mg/L; 1 means 10 mg/L. The dots reflect the locations of the springs from which analyses have been used in the calculations of the maps. Scales on the x- and y-axis are in km. The maximum value of 2.4 equal to 25 mg/L, the lowest values (below -2.0) indicate regions with less than 0.01 mg/L aluminium in the water.

increase the solubility of aluminium by forming strong $AlSO_4^+$ complexes (HEM 1968), 8.00cmwhich may be the reason for increased aluminium concentrations in the Western Meso-Cenozoic Basin.

In this study, the number of Fe^{2+} analyses is much higher than the number of Al^{3+} analyses. Except for the same acid springs where aluminium has high concentrations, relatively high concentrations of iron (up to 6 mg/L) are found in the Western Meso-Cenozoic Basin (*Map 13*). These moderately high Fe concentrations found in thermal and mineral waters located in this part of the country could be mainly related to a limestone dissolution process (limestone being the most common geological formation in this region), which gives rise to the formation of *"terra rossa"*. The so-called *"terra rossa"* is a residual soil (a detrital deposit resulting from the accumulation of clays, fine sands and iron oxides) due to the dissolution of limestone. This way, most of the iron present in the limestone formations will be retained in the residual product of the alteration process.

Photo 6: General view of the Carlão thermal region (district of Vila Real).

Another region with significant Fe^{2+} concentrations in freshwater springs is found in the region north of the towns of Loulé and Tavira in the Algarve region (in the Southern Meso-Cenozoic Basin). These waters issue from rocks of the Schist-Greywacke Complex from Serra do Caldeirão (part of the Iberian Massif), where they receive their Fe^{2+} present in the clastic minerals such as pyroxene (augite) and mica (biotite). This Fe^{2+} is quickly precipitated after the discharge of the waters (D´ALMEIDA & DE ALMEIDA 1966).

Iron (log of mg/L)

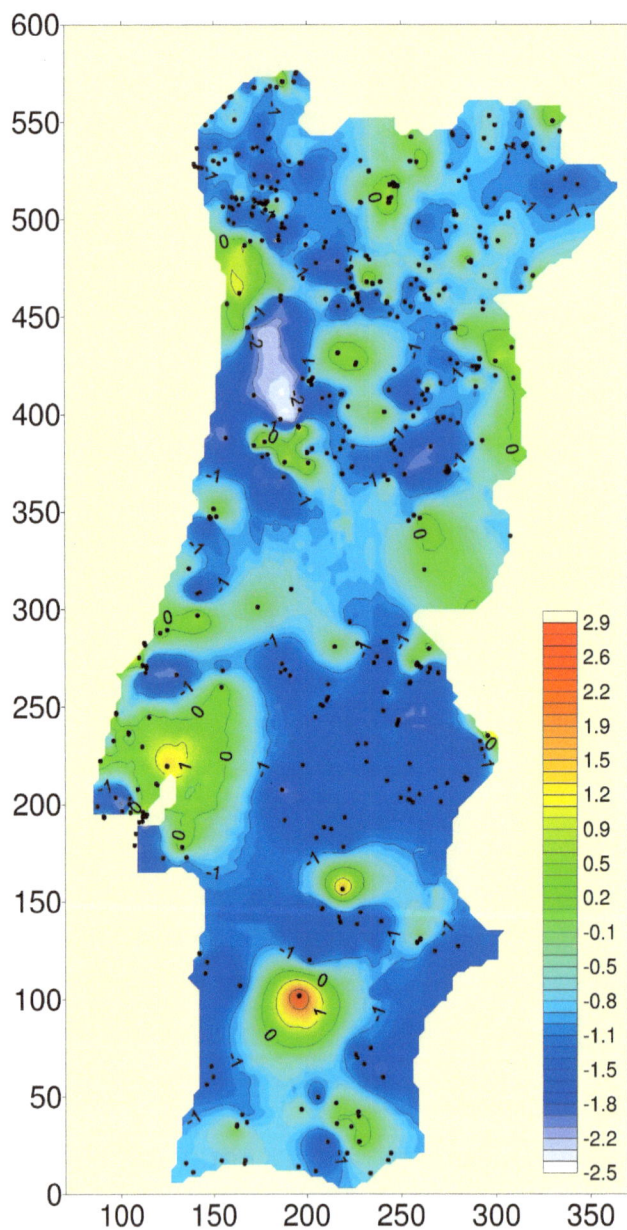

Map 13: Distribution of Fe²⁺ in thermal and mineral waters from the Portuguese mainland. The numbers on the lines in the Map indicate the logarithm of the concentrations in mg/L (i.e. 1 means 10 mg/L; 2 means 100 mg/L. The dots reflect the locations of the springs from which analyses have been found and used in the calculations of the maps. Scales on the x- and y-axis are in km. The maximum value of 2.9 equals to 790 mg/L, the lowest values (below -2.0) indicate regions with less than 0.01 mg/L iron in the water.

32

Species like HS⁻, F⁻ and Li⁺ are very important in natural waters as they have very specific influences on human health. Waters containing HS⁻ are recommended in medical hydrology for the treatment of respiratory (sinusitis, chronic rhinitis, laryngitis and bronchitis among others), dermatological, rheumatic and muscular-skeletal pathologies (rheumatoid arthritis, post-traumatic squeals) applied in many Portuguese spas (thermal health centres such as São Pedro do Sul, Caldas da Rainha, Caldas do Gerês, Cabeço de Vide, Carlão, S. Lourenço among others (*e.g.* FERREIRA GOMES *et al.* 2001; MARQUES *et al.* 1999b; 2000; 2008; 2010).

H₂S gas which escapes from solutions containing the HS⁻ ion is a toxic gas with a very specific smell of rotten eggs. Its maximum allowable concentration in the EU is 10 ppm, and its LC50 for rats is about 500 mg/L. Hydrogen sulphide is a mitochondrial poison. Its action on mitochondria is similar to that of cyanide through inhibition of cytochrome oxidase (iron containing protein). In addition, hydrogen sulphide binds to hemoglobin in red blood cells interfering with oxygen transport. Exposure to H₂S occurs primarily by inhalation but can also occur by ingestion (contaminated food) and skin (water and air). Once taken into the body, it is rapidly distributed to various organs, including the central nervous system, lungs, liver, muscle, etc. This is why in Portuguese thermal centres the administered patient doses are very small: 40-50 ml, 2 to 3 times in the morning 30 minutes apart (DIEGUES & MARTINS 2001). Hydrogen sulphide-rich waters are among the oldest used as Spas in Portugal (see *e.g.* MARQUES *et al.* 2010).

Photo 7: The sulphide rich spring at São Lourenço (district of Bragança) with the saint who protects the users above the outlet.

This species is found in two general geologies (*Map 14*): one is related to the granitic rocks at the northern part of the Iberian Massif, as well as in the Serra de Monchique in the southern Algarve region. The other is related to

Bisulphide (log of mg/L)

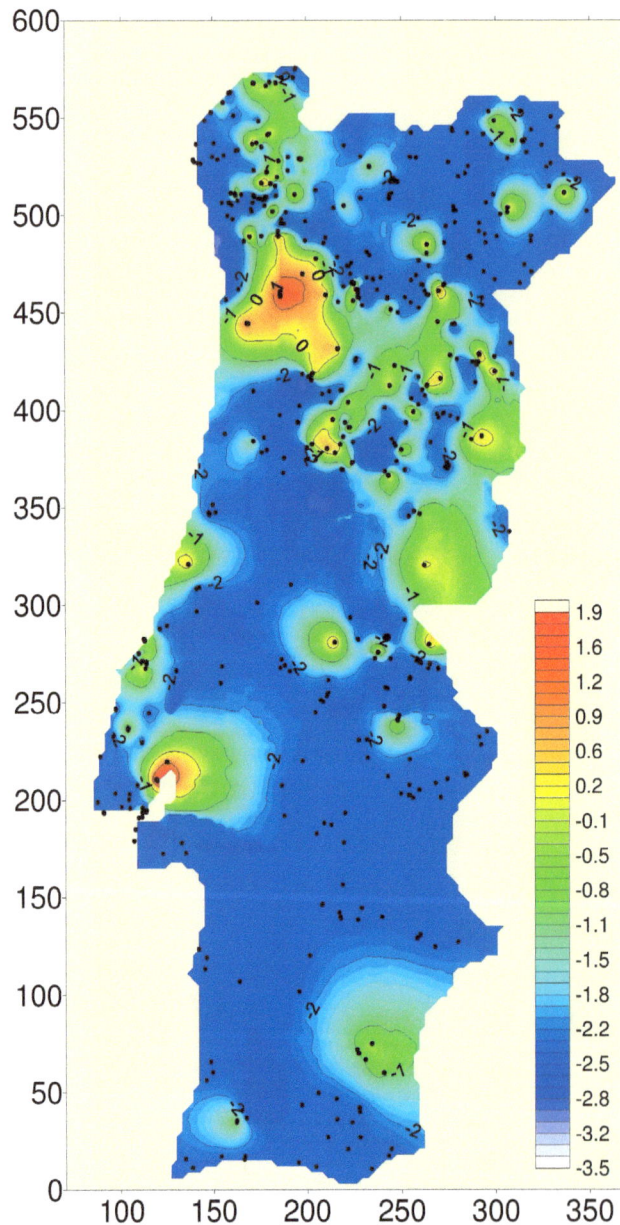

Map 14: Distribution of HS⁻ in thermal and mineral waters from the Portuguese mainland. The numbers on the lines in the Map indicate the logarithm of the concentrations in mg/L (i.e. 0 means 1 mg/L; 1 means 10 mg/L. The dots reflect the locations of the springs from which analyses have been used in the calculations of the maps. Scales on the x- and y-axis are in km. The maximum value of 1.9 equals to 79 mg/L, the lowest values (below -2.0) indicate regions with less than 0.01 mg/L bisulphide in the water.

34

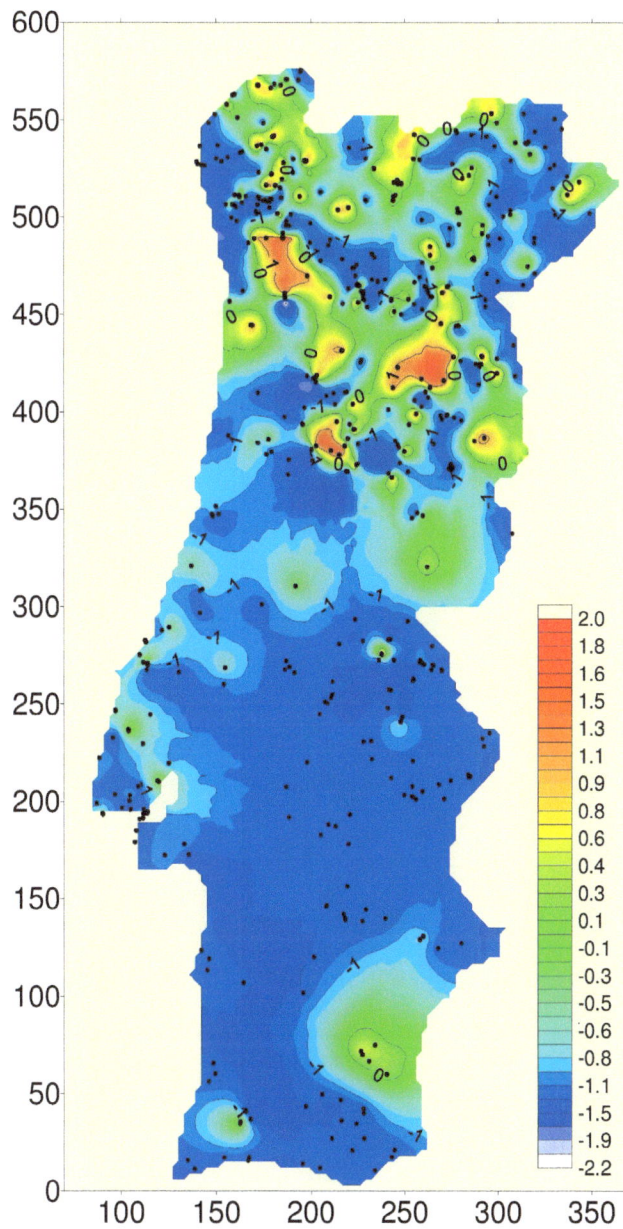

Fluoride (log of mg/L)

Map 15: Distribution of F⁻ in thermal and mineral waters from the Portuguese mainland. The numbers on the lines in the Map indicate the logarithms of the concentrations in mg/L (i.e. 0 means 1 mg/L; 1 means 10 mg/L. The dots reflect the locations of the springs from which analyses have been used in the calculations of the maps. Scales on the x- and y-axis are in km. The maximum value of 2.0 equals to 100 mg/L, the lowest values (below -2.0) indicate regions with less than 0.01 mg/L fluoride in the water.

35

Mesozoic sediments in the Western Meso-Cenozoic Basin, and in a small area related to Paleozoic sediments in the south-east (CALADO & CHAMBEL 1999).

In these areas it is suggested that HS⁻ originates in the mineral waters due to bacterial reduction of sulphates present in gypsum deposits in the Western Basin and possibly paleo seawater in the south-eastern region (CALADO & CHAMBEL 1999). However, recent studies to detect bacteria in sulphurous water from Caldas da Rainha failed to find any of them (HENRIQUE GRAÇA, *pers. comm.*). Research presented by HOOPER *et al.* (2010) showed that iron reducing bacteria are able of using sulphur in gypsum as an energy source, producing hydrogen sulphide and other reduced sulphur compounds. However as no bacteria have been found in any of the sulphide rich mineral and thermal waters in the Portuguese mainland yet, the origin of these waters is still unknown.

Photo 8: The Spa building at Monção (district of Viana do Castelo), one of the famous fluoride rich spas in Portugal.

Fluoride is also a species with very important health properties. Medical advice adopted and recommended by the WHO (2004) is, respectively, that in drinking water concentrations between 0.5 and 1 mg/L are optimal and concentrations above 1.5 mg/L are considered to be detrimental to health. Higher fluoride doses have been linked to the development of dental fluorosis (mottled enamel and decay) and in extreme cases, skeletal fluorosis (osteosclerosis, bone deformation and debilitation) (SHARMA 2003; ALVAREZ *et al.* 2009). Fluoride concentrations in groundwater are in most of the cases high in the northern part of the country (*Map 15*) due to the presence of calc-alkaline granites of Paleozoic age and F-bearing minerals such as fluorapatite ($Ca_5(PO_4)_3F$) and fluorite (CaF_2) mainly present in hydrothermal vein deposits (DIAS *et al.* 1998). Calcium controls the F concentration in thermal and mineral waters due to the low CaF_2 solubility product ($Ks_{CaF2} =$

$10^{-10.57}$, *e.g.* M̶ARQUES *et al.* 2003). The F⁻ concentration can become higher, up to 21 mg/L, if Ca does not easily go into solution, for example due to precipitation of $CaCO_3$ at high pH as many F-rich waters have a pH that can be as high as 9, or because of the presence of calcium poor granites (A̶BDELGAWED *et al.* 2009).

Lithium is an ion with very pronounced pharmaceutical properties. A number of lithium salts are used as mood stabilizing drugs in medicine, primarily in the treatment of bipolar disorder, where they have a role in the

Photo 9: Vidago Areal 3, a lithium rich artesian well near Vidago (district of Vila Real).

treatment of depressions and particularly of mania, both acutely and in the long term. Upon ingestion, lithium becomes widely distributed in the central nervous system and interacts with a number of neurotransmitters and receptors, decreasing norepinephrine release and increasing serotonin synthesis. According to F̶ORNAI *et al.* (2008), M̶ARMOL (2008) and Q̶UIROZ *et al.* (2010) it also appears to delay the progression of neurodegenerative diseases such as Parkinson's disease, Alzheimer's disease and Amyotrophic Lateral Sclerosis (ALS). Two studies, one in Japan (O̶GHAMI *et al.* 2009) and another in the USA (S̶CHRAUZER & S̶HRESTHA 1990), showed that lithium contents in drinking water could have beneficial effects for society: communities with higher amounts of lithium in their drinking water (*e.g.* 70-170 µg/L) had significantly lower suicide rates and violent crimes as homicide and rape than communities with lower levels.

Knowledge of the Li⁺ concentration in water thus is of very high importance. Unfortunately, in our database Li is only known in a small minority of samples. From the available data it can be concluded that Li is present in

Lithium (log of mg/L)

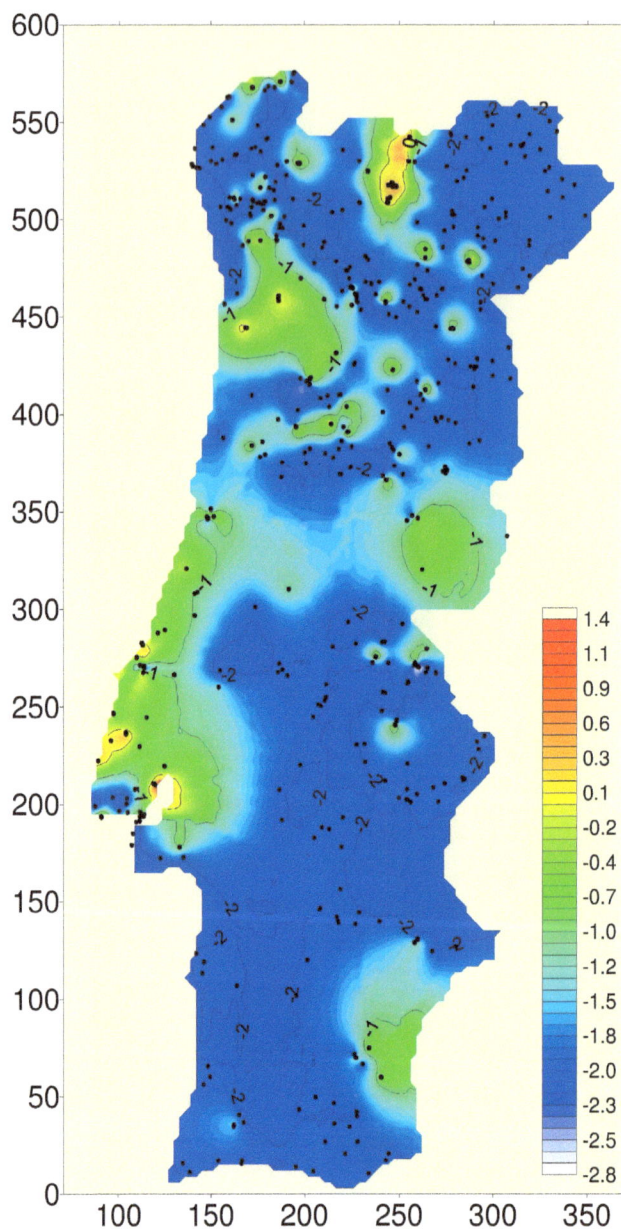

Map 16: Distribution of Li⁺ in thermal and mineral waters from the Portuguese mainland. The numbers on the lines in the Map indicate the logarithms of the concentrations in mg/L (i.e. 1 means 10 mg/L; 2 means 100 mg/L). The dots reflect the locations of the springs from which analyses have been used in the calculations of the maps. Scales on the x- and y-axis are in km. The maximum value of 1.4 equals to 25 mg/L, the lowest values (below -2.0) indicate regions with less than 0.01 mg/L lithium in the water.

the northern part of the country in areas mainly dominated by granitic rocks where Li^+ is present in pegmatite minerals as spodumene ($LiAlSi_2O_6$), amblygonite ((Li, Na)$Al(PO_4)(F,OH)$), litiophylite ($LiMnPO_4$) and lepidolite (($K(Li,Al)_3(Si,Al)_4O_{10}(F,OH)_2$). The areas of Guarda, Viseu, Vila Real and Viana do Castelo have the highest potential for lithium exploitation. However, highest Li concentrations in mineral and thermal waters are found in the Na-HCO_3-CO_2-rich thermal and mineral waters issuing along the Penacova-Régua-Verín fault such as Vilarelho da Raia, Chaves, Vidago and Pedras Salgadas (*Map 16*).

Photo 10: The "buvette" in Pedras Salgadas (district of Vila Real). Pedras Salgadas is one of the most lithium rich mineral waters in Portugal.

Fluoride in these waters is also relatively high (although not as high as in the HS⁻-rich areas). Lithium here should be dissolved from Li-rich minerals in the rock due to reaction with upper-mantle CO_2. AIRES-BARROS *et al.* (1998) indicated that the constant ratios of the major ionic species HCO_3^-, Na^+ and Li^+, when plotted against a conservative element such as Cl⁻ indicates water-rock interaction related with the same geological (granitic) environment, where higher salinities should correspond to larger residence times probably associated with deeper circulations. This trend is more evident in a Li^+ vs. Cl⁻ plot where the typically conservative (mobile) Cl⁻ is compared with another rather conservative species such as Li^+ (AIRES-BARROS *et al.* 1998).

The other area with relatively high Li^+ concentrations is the Western Meso-Cenozoic Basin, where Li^+ enters the thermal and mineral waters due to dissolution of relatively Li-rich salt deposits in that area (*Map 16*).

It is clear that there is some overlap between these three elements in regions with magmatic rocks, where HS⁻ could have formed. In the other

39

regions these similarities are not found, for example the CO_2-rich waters in the northern part of the country contain no HS^-.

Considering the importance of this type of elements that have clear potential effects on human health it is recommended that in future studies these effects of minor and trace elements are taken into account. Medical geochemical studies of the thermal and mineral waters in Portugal would be recommended to understand further the effects of population exposure to these elements.

Comparison with other geochemical mapping studies

The only other published geochemical mapping studies applied to Portugal are as previously reported the Soil Geochemical Atlas of Portugal (INÁCIO *et al.* 2008) and the Europa wide FOREGS low density geochemical mapping (SALMINEN *et al.* 2005).

Two of the maps presented by INÁCIO *et al.* (2008) show elements also covered in this study. Soil geochemical maps of Ca and Al are presented, with a few other elements. The comparison of those two maps with our Ca and Al distributions puts in evidence very clear differences, which are due to the very different geochemical properties of these elements. Since Ca is a relatively soluble element, its distribution in thermal and mineral waters is fairly comparable to the Ca distribution in soils in Portugal, with high values in the Western and Southern Meso-Cenozoic Basins and much lower values in the northern part of the Iberian Massif. Concerning Al the behaviour is completely different. At most pH values this element presents a very low solubility (APPELO & POSTMA 2005). Al is soluble as Al^{3+} or $AlOH^{2+}$ only under acidic conditions, and moderately as fluoride and sulphate complexes at more neutral pHs. On the other hand, in rocks Al is a major element and the soil geochemical map shows the total Al concentration, and not the available fraction of this chemical element that is soluble in water. It shows high values in areas where granitic rocks are the predominant rock type and lower values elsewhere, including very low values in the Western Meso-Cenozoic Basin where sediments are found, including limestones which contain only very little Al. On the other hand , in the thermal and mineral water map, it is shown that in most of the Portuguese mainland, Al has a low to very low concentration (< 0.02 mg/L), showing high values in a few particular spots (up to 200 mg/L) which are the result from individual discharges with very low pH waters, mostly the result from "*acid drainage*" environments, one of them related with mining activities in the Iberian Pyrite Belt. As such, these two elements show the extreme differences between a water-related and soil related map, which are sampled on the same scale. The more soluble elements will show more similarities between these two types of maps, while less soluble elements (or elements which are

40

only soluble in specific environments) will show very different distributions.

Although we can only compare two elements with the Soil Geochemical Atlas, we can compare most elements using the information obtained from the FOREGS low density geochemical mapping of Europe (SALMINEN *et al.* 2005). However, as this was a low density study (on average one sample per 4700 km^2) it is sampled with a much lower density than both the Soil Geochemical Atlas (1 sample per 135 km^2) and the thermal and mineral water mapping (average 1 sample per 200 km^2) studies. So, it will not be possible to compare on a bigger than the low density scale. However, it would be interesting to summarise comparisons between the thermal and mineral water geochemical maps and the different FOREGS maps.

Photo 11: The building where the "buvette" of the Chaves Thermal Area (district of Vila Real) is located. The Chaves Thermal Area supplies the hottest (76 ºC) mineral water from Mainland Portugal. The highest temperature is encountered in borehole AC2 which is drilled a few tens of meters from this building.

Sodium is partly a very soluble element, but also available at high concentrations in rocks, mainly in feldspars. Comparing FOREGS stream sediment and stream water maps with the thermal and mineral water maps, we believe are most representative for the geochemistry of fresh rocks, already shows nice differences for Na. Stream waters shows a gradient from high to low values going south to north, a trend which represents most probably the climatic difference between south, which is dry, and the north which has much more precipitation. The stream3.80cm sediments containing of relatively fresh rock fragments show a more diverse variation with higher values in predominately granitic areas and lower values elsewhere. Both maps are very different from the thermal and mineral water map, which shows very high Na concentrations in areas with shallow salt (NaCl) deposits, low values in most of Portuguese mainland, and the lowest values, at the central part of the Iberian Massif, associated with schistose

41

rocks from the Schist-Greywacke Complex. This is a clear example of the different sampling approach that is obtained using thermal and mineral water mapping.

Hydrogen carbonate ion in stream water also shows a striking difference: in stream waters again we see a decrease in concentration going south to north. This is the result of more common carbonate rocks in the south, but partly also due to the dryer climate in the south. This characteristic is partly also visible in the thermal and mineral water map, with lower values in the north than in the south. However, the very high values found in areas near very deep upper mantle CO_2 degassing faults (MARQUES et al. 1998; 2001; 2006; CARREIRA et al. 2010), in the northern part of Portuguese mainland, are not visible on a stream water map. As such, it shows that the information available in the FOREGS mapping represents surface geochemistry when looking at sediments solely, and this might be possibly coupled with climatic and even human effects when surface waters are taken into account.

CONCLUSIONS

During the past 150 years a very large amount of chemical data, especially for major elements, from the thermal and mineral waters on the Portuguese Mainland have been analysed. From this large amount of data geochemical maps from Mainland Portugal were produced that represent chemical data of over 500 locations with thermal and mineral waters. As no new analyses were made, all data used in this study were obtained from the literature. An advantage of this approach is that no field work campaign had to be set up, a disadvantage may be that the quality of the analyses is not in all cases, especially the older analyses, really well known. However, as analytical data from individual springs that were analysed more than once, often decades apart, showed very comparable results it is assumed for this study that the analytical quality of most analyses is very good or at least acceptable. This approach was used to test if it could lead to meaningful geochemical maps. The maps of the chemical species Li^+, Na^+, Mg^{2+}, Ca^{2+}, Al^{3+}, Fe^{2+}, F^-, Cl^-, SO_4^{2-}, NO_3^-, HCO_3^-, HS^- and SiO_2, as well as the total dissolved solids and the pH can, with the possible exception of nitrate, be well explained when compared with the known geology and geochemistry of the country. It is shown that the extra information that can be obtained from a thermal and mineral water geochemical mapping study as opposed to more *"conventional"* sediment and stream water mapping gives a good representation of the deeper geochemistry of the mapped area. While the more conventional techniques result in sampling of just the upper few meters of the geology, thermal and mineral waters generally circulate much deeper, up to about four kilometres (*e.g.* MARQUES et al. 2001) and as such their geochemistry is representative not only for a certain area, but also for the volume of rock below that area, potentially giving information on the upper few kilometres of the upper crust, although it needs to be realised that in some areas the circulation will be much shallower. This information not otherwise available is very useful when the geology and geochemistry in different areas are compared.

Despite the advantages that this extra information can give, there are still some disadvantages in using thermal and mineral waters for geochemical mapping. Most importantly sampling points cannot be chosen randomly. Obviously only thermal and mineral waters can be taken and their distribution over a certain area can differ vastly, as is already shown over the relatively small area of the Portuguese mainland. Except for this it is also not precisely known what will be the area and depth that is actually sampled by a specific water sample. Conceptual models have only been developed for the most important thermal and mineral waters (e.g. MARQUES et al. 2001; 2003; 2006; 2008; 2010; CARREIRA et al. 2010). In areas with a very high sample density the sampling zones could be much smaller than in areas with a lower density, and sampling zones might overlap. Another risk is that some waters will have a relatively shallow circulation, introducing the risk for agricultural contamination, which may be the reason for increased nitrate concentrations in several thermal and mineral waters (mainly at springs, much less in boreholes).

In spite of these potential problems the produced maps show in general that the distributions of the chemical elements reflect very nicely a combination of the geochemical composition of the underground with important chemical processes that take place, indicating the potential usefulness of this approach to make geochemical maps.

OUTLOOK

One of the key requests for the effective study of thermal and mineral waters geochemistry in relation to Geology and Health is the production of multi-element Atlases showing the distribution of the elements on a regional scale. The selected method for assembling such Atlas can vary according to a number of geological, geochemical and environmental factors. However, the fundamental consideration is to assist (combining other relevant sources of information) in defining, quickly and inexpensively, potential problematic areas. Many sampling and analytical techniques can be employed. Each technique and approach has its own prospective, constraints and interpretation problems. Whatever method is chosen, the use of computer-based statistical data reduction, analysis and map gathering is required.

Mapping geochemical signatures from the wide variety of thermal and

43

mineral waters found on the Portuguese mainland showed that the distribution of most major elements represent a combination of the geochemical composition of the upper layer of the crust and the important geochemical processes that take place there. This methodological approach (including generally a deeper sampling) also reveals considerable discrepancies when compared with more conventional sampling techniques such as stream water or stream sediments. As data were taken from published and partially older data only major elements could be assessed into much detail. It is shown that these results are encouraging and it is recommended that the thermal and mineral springs should be re-sampled countrywide and that a full analytical program (including minor and trace elements) should be applied. This way the wealth of information stored in the unique collection of Portuguese thermal and mineral waters could finally be utilised to its full extent.

Considering the importance of the elements that might have clear positive or negative effects on human health, it is recommended that in future studies these effects of minor and trace elements be also taken into account. Medical geochemical studies of the thermal and mineral waters in Portugal would be recommended to understand further not only the beneficial but also the more dangerous effects of the population exposure to some chemical species.

Acknowledgements - This book was originally supported by the Centro de Petrologia e Geoquímica of Instituto Superior Técnico (CEPGIST) that was merged into the Centro de Recursos Naturais e Ambiente (CERENA) of the Instituto Superior Técnico (Lisbon) in 2014. Part of the contribution of HGME was funded by the Portuguese Foundation for Science and Technology (FCT) under the Ciência 2007 Program. *Map 1* was improved by José Teixeira and Helder Chaminé and we thank them both. *Photos 1, 2, 6, 7, 8, 9* and *11* have been taken by José Manuel Marques, *photos 4, 5* and *10* have been taken by Paula Carreira, *photo 3* has been taken by Amélia Dionisio. All photos have been used by permission. We would like to express our gratitude to Prof. Luís Aires-Barros, President of the Academy of Sciences of Lisbon for writing the preface to this monograph.

REFERENCES

ABDELGAWAD AM, WATANABE K, TAKEUCHI S, MIZUNO T (2009) The origin of fluoride-rich groundwater in Mizunami area, Japan – Mineralogy and geochemistry implications. Eng Geol 108:76-85

ACCIAIUOLI LMC (1952) Le Portugal hydrominéral. I Volume. DGM Lisbonne. 284 pp

ACCIAIUOLI LMC (1953) Le Portugal hydromineral. II Volume. DGM Lisbonne. 290 pp

AIRES-BARROS L (1979) Notas sobre a geoquímica das águas minerais portuguesas. Boletim do Museu, Laboratório de Mineralogia e Geologia do Faculdade de Ciências de Lisboa 16:123-135

AIRES-BARROS L (1989) Geothermal Resources in Portugal. Anais da UTAD 2:11-22

AIRES-BARROS L, MARQUES JM (2000) Portugal Country Update. Proceedings of the World Geothermal Congress (Iglesias E, Blackwell D, Hunt T, Lund J, Tamanyu S eds.), Kyusu – Tohoku, Japan, 39-44

AIRES-BARROS L, MARQUES JM, GRAÇA RC (1995) Elemental and isotopic geochemistry in the hydrothermal area of Chaves, Vila-Pouca-de-Aguiar (northern Portugal). Environ Geol 25:232-238

AIRES-BARROS L, MARQUES JM, GRAÇA RC, MATIAS MJ, VAN DER WEIJDEN CH, KREULEN R, EGGENKAMP HGM (1998) Hot and cold CO_2-rich mineral waters in Chaves geothermal area (Northern Portugal). Geothermics 27:89-107

ALBU M, BANKS D, NASH H (1997) Mineral and thermal groundwater resources. Chapman and Hall, London, UK.

ALVAREZ JA, RESENDE KMPC, MAROCHO SMS, ALVES FBT, CELIBERTI P, CIAMPONI AL (2009) Dental fluorosis: Exposure, prevention and management. J Clin Exp Dent 1:14-18

ANTUNES DA SILVA MPT (2007) O director técnico na gestão de concessões de água gasocarbónica. O valor acrescentado das Ciências da Terra no termalismo e no engarrafamento da água. Artigos seleccionados do II Fórum Ibérico de Águas Engarrafadas e Termalismo, Porto 22 a 24 de Novembro de 2006,129-139

APPELO CAJ, POSTMA D (2005) Geochemistry, groundwater and pollution. 2nd edition. CRC Press, Taylor & Francis Group, Boca Raton

ARNÓRSSON S, GUNNLAUGSSON E, SVAVARSSON H (1983) The chemistry of geothermal waters in Iceland. III. Chemical geothermometry in geothermal investigations. Geochim Cosmochim Acta, 47:567-577

AZARÊDO AC, WRIGHT VP, RAMALHO MM (2002) The middle-late Jurassic forced regression and disconformity in central Portugal: eustatic, tectonic and climatic effects on a carbonate ramp system. Sedimentology 49:1339-1370

BASTOS C (2008) O Novo Aquilégio. Published on the internet: http://www.aguas.ics.ul.pt/. Last visited 9 December 2014

BARRIGA FJAS, CARVALHO D (1983) Carboniferous volcanogenic sulphide mineralizations in South Portugal (Iberian Pyrite Belt). In: Lemos de Sousa MJ, Oliveira JT, eds. The Carboniferous of Portugal. Memórias dos Serviços

Geológicos de Portugal, 29: 99-116

CABEÇAS R, CARVALHO JM, NUNES JC (2010) Portugal Country Geothermal Update 2010, Proceedings World Geothermal Congress 2010, Bali, Indonesia, pp 1-9

CABRAL J (1989) An example of intraplate neotectonic activity Vilariça Basin, Northeast Portugal. Tectonics 8:285-303

CALADO C, CHAMBEL A (1999) Un unexpected mineral sulphide water type in the Iberian Pyrite Belt (South Portugal). XXIX IAH Congress, Hydrogeology and Land Use Management, Miriam Fendeková & Marián Fendek Eds, Bratislava, Slovak Republic, pp. 671-676

CARNEIRO MC (1993) As Caldas de Chaves – Fonte de Energia / Geothermia. Chaves, Portugal.

CARREIRA PM, MARQUES JM, ANDRADE M, MATIAS H, LUZIO R, MONTEIRO SANTOS F, NUNES D (2004) Isotopic, geochemical and geophysical studies to improve Caldas de Monção thermomineral waters conceptual circulation model (NW Portugal). Cadernos Laboratorio Xeolóxico de Laxe 29:147-170.

CARREIRA PM, MARQUES JM, CARVALHO MR, CAPASSO G, GRASSA F (2010) Mantle-derived carbon in Hercynian granites. Stable isotopes signatures and C/He associations in the thermomineral waters, N-Portugal. J Volcanol Geoth Res 189:49-56

COSTA IR, BARRIGA F, MATA J, MUNHÁ JM (1993) Rodingitization and serpentinization processes in Alter-do-Chão Massif (NE Alentejo). Proceedings of the IX Semana de Geoquímica (Noronha F, Marques M and Nogueira P., eds.). Universidade do Porto. Faculdade de Ciências. Museu e Laboratório Mineralógico e Geológico 27-31

CRUZ JV, COUTINHO RM, CARVALHO MR, OSKARSSON N, GISLASON SR (1999) Chemistry of waters from Furnas volcano, São Miguel, Azores: fluxes of volcanic carbon dioxide and leached material. J Volcanol Geoth Res 92:151-167

DA COSTA AM, FRANCÉ AP, FERNANDES J, LOURENÇO MC, MIDÕES C, RIBEIRO LF, OLIVEIRA E (2006) Estudo Hidrogeoquímico do Sistema Aquífero Moura-Ficalho. Actas/CD do 8º Congresso da Água - "Água - Sede de Sustentabilidade", Figueira da Foz, 13-17 de Março de 2006, 20 pp

D´ALMEIDA A, DE ALMEIDA J (1966) Inventário Hidrológico de Portugal. 1º Volume. Algarve. Instituto de Hidrologia de Lisboa

D´ALMEIDA A, DE ALMEIDA J (1970) Inventário Hidrológico de Portugal. 2º Volume. Trás-os-Montes e Alto Douro. Instituto de Hidrologia de Lisboa

D´ALMEIDA A, DE ALMEIDA J (1975) Inventário Hidrológico de Portugal. 3º Volume. Beira Alta. Instituto de Hidrologia de Lisboa

D´ALMEIDA A, DE ALMEIDA J (1988) Inventário Hidrológico de Portugal. 4º

Volume. Minho. Instituto de Hidrologia de Lisboa

DALMEYER RD, MIRTÍNEZ GARCÍA E (1990) Pre-Mesozoic geology of Iberia. Springer Verlag

DARNLEY AG, GARRETT RG (1990) International geochemical mapping. J Geochem Expl 49:1-250

DARNLEY AG, BJÖRKLUND A, BÖLVIKEN B, GUSTAVSSON N, KOVAL PV, PLANT JA, STEENFELT A, TAUCHID M, XUEJING X, GARRETT RG, HALL GEM (1995) A global geochemical database for environmental and resource management. Recommendations for international geochemical mapping. Final report of IGCP project 259. Unesco Publishing

DIAS G, LETERRIER J, MENDES A, SIMÕES PP, BERTRAND JM (1998) U-Pb zircon and monazite geochronology of post-collisional Hercynian granitoids from the Central Iberian Zone (Northern Portugal). Lithos 45:349-369

DIEGUES P, MARTINS V (2001) Águas termais: riscos e benefícios para a saúde. Encontro Técnico Água e Saúde. Caparica, Portugal

DGGM [DIRECÇÃO-GERAL DE GEOLOGIA E MINAS] (1992) Termas e águas engarrafadas em Portugal. Lisboa

EGGENKAMP HGM & MARQUES JM (2013) A comparison of mineral water classification techniques: Occurrence and distribution of different water types in Portugal (including Madeira and the Azores). J Geochem Expl 132:125-139

EGGENKAMP HGM, MARQUES JM, GRAÇA H (2010) Fractionation of Cl isotopes during precipitation of NaCl from a nearly pure NaCl brine. Geochim Cosmochim Acta 74 (Supp.1): A260

EGGENKAMP HGM, MARQUES JM, GRAÇA H (2013) Application of stable chlorine isotopes to develop a conceptual model for the origin of the ground water circulating near the "salinas" at Rio Maior (Portugal). Com Geol 100(1):49-53

ESPAÑA JS, PAMO LE, SANTOFIMIA E, ADUVIRE O, REYES J, BARETTINO D (2005) Acid mine drainage in the Iberian Pyrite Belt (Odiel river watershed, Huelva, SW Spain): Geochemistry, mineralogy and environmental implications. Appl Geochem 20:1320-1356

ERSHA [ESTUDO DOS RECURSOS HÍDRICOS SUBTERRÂNEOS DO ALENTEJO] (2000) Nascentes do Alentejo. Centro de Geologia da Universidade de Lisboa and Instituto da Água, Lisboa

FERREIRA GOMES LM, AFONSO DE ALBUQUERQUE FJ, FRESCO H (2001) Protection areas of the São Pedro do Sul Spa, Portugal. Engin Geol 60:341-349

FORNAI F, LONGONE P, CAFARO L, KASTSIUCHENKA O, FERRUCCI M, MANCA ML, LAZZERI G, SPALLONI A, BELLIO N, LENZI P, MODUGNO N, SICILIANO G, ISIDORO

C, Murri L, Ruggieri S, Paparelli A (2008) Lithium delays progression of amyotrophic lateral sclerosis. Proc Natl Acad Sci USA 105:2052-2057.

Galego Fernandes P, Carreira P, Oliveira da Silva M (2008) Anthropogenic sources of contamination recognition – Sines coastal aquifer (SW Portugal). J Geochem Expl 98:1-14

Garret RG, Reimann C, Smith DB, Xie X (2008) From geochemical prospecting to international geochemical mapping: a historical overview. Geochemistry: Exploration, Environment, Analysis 8:202-217.

Greber E (1994) Deep circulation of CO_2-rich palaeowaters in a seismically active zone (Kuzuluk/Adaparazi, northwestern Turkey). Geothermics 23:151-174.

Hawkes HE (1976) The early days of exploration geochemistry. J Geochem Expl 6:1-11.

Hem JD (1968) Graphical methods for studies of aqueous aluminum hydroxide, fluoride, and sulphate complexes. USGS Water-Supply Paper 1827-B, 33 pp.

Henriques FF (1728) Aquilegio medicinal. Officina Musica, Lisboa.

Harrington LF, Cooper EM, Vasudevan D (2003) Fluoride sorption and associated aluminum release in variable charge soils. J Coll Interf Sc 267:302-313.

Hooper DG, Shane J, Straus DC, Kilburn KH, Bolton V, Sutton JS Guilford FT (2010) Isolation of sulfur reducing bacteria and oxidizing bacteria found in contaminated drywall. Int J Mol Sci 11:647-655.

Inácio M, Pereira V Pinto M (2008) The soil geochemical atlas of Portugal: Overview and applications. J Geochem Expl 98:22-33.

INE [Instituto Nacional de Estatística] (2014) Census 2011 http://censos.ine.pt/xportal/xmain?
xpid=CENSOS&xpgid=censos2011_apresentacao

Lepierre Ch (1921) Un nouveau type d´eaux minérales: les eaux nitratées. Comptes rendus 173:783-786.

Lepierre Ch, Herculano de Carvalho A (1930) Les eaux radioactivités de Caria (Beira Baixa, Portugal). XII Congres International d´Hydrologie, Lisbonne.

L.N.E.G. [Laboratoria Nacional de Energia e Geologia] (2010) Carta Geologia de Portugal (1:1.000.000).

Lopo Mendonça J, Oliveira da Silva M, Bahir M (2004) Considerations concerning the origin of the Estoril (Portugal) thermal water. Estudios Geol 60:153-159.

Lourenço C, Cruz J (2005) Aproveitamentos geotérmicos em Portugal

Continental. XV Encontro Nacional do Colégio de Engenharia Geológica e de Minas da Ordem dos Engenheiros. Ponta Delgada, pp. 1-9.

LUZES O, NARCISO A, LEPIERRE CH., L'ARROCHELLA C, DURÃO F (1930) Le Portugal hydrologique et climatique. Deuxième Partie. DGM et IH, Lisbonne.

LUZES O, NARCISO A, LEPIERRE CH., L'ARROCHELLA C, DURÃO F (1934) Le Portugal hydrologique et climatique. Troisième Partie. DGM et IH, Lisbonne.

LUZES O, NARCISO A, LEPIERRE CH., L'ARROCHELLA C, DURÃO F (1935) Le Portugal hydrologique et climatique. Quatrième Partie. DGM et IH, Lisbonne.

MARMOL F (2008) Lithium: Bipolar disorder and neurodegenerative diseases. Possible cellular mechanisms on the therapeutic effects of lithum. Prog Neuro-Psychopharma Biol Psych 32:1761-1771.

MARQUES JM, CARREIRA PM, AIRES-BARROS L, GRAÇA RC (1998) About the origin of CO_2 in some HCO_3/Na/CO_2-rich Portuguese mineral waters: Geoth Res Coun Trans 22:113-117.

MARQUES JM, AIRES-BARROS L, GRAÇA RC (1999a) Geochemical and isotopic features of hot and cold CO2-rich mineral waters of northern Portugal: a review and reinterpretation. Bull Hydrogéo 17:175-183.

MARQUES JM, AIRES-BARROS L, GRAÇA RC (1999b) Isotopic and chemical signatures of low-temperature sulphurous mineral waters (northern Portugal): preliminary results. Geoth Res Council Trans 23:327-332.

MARQUES JM, AIRES-BARROS L, GRAÇA RC (2000) Genesis of low-temperature sulphurous mineral waters (Northern Portugal): a geochemical and isotopic approach. Proceedings of the World Geothermal Congress (Iglesias, E., Blackwell, D., Hunt, T. Lund, J. & Tamanyu, S. eds.), Kyusu – Tohoku, Japan, 1407-1412.

MARQUES JM, MONTEIRO SANTOS FA, GRAÇA RC, CASTRO R, AIRES-BARROS L, MENDES VICTOR LA (2001) A geochemical and geophysical approach to derive a conceptual circulation model of CO2-rich mineral waters: A case study of Vilarelho da Raia, northern Portugal. Hydrogeol J 9:584-596.

MARQUES JM, ESPINHA MARQUES J, CARREIRA PM, GRACA RC, AIRES-BARROS L, CARVALHO JM, CHAMINE HI, BORGES FS (2003) Geothermal fluids circulation at Caldas do Moledo area, Northern Portugal: geochemical and isotopic signatures. Geofluids 3:189-201.

MARQUES JM, ANDRADE M, CARREIRA PM, EGGENKAMP HGM, GRAÇA RC, AIRES-BARROS L, ANTUNES DA SILVA M (2006) Chemical and isotopic signatures of Na/HCO_3/CO_2-rich geofluids, North Portugal. Geofluids 6:273-287.

MARQUES JM, CARREIRA PM, CARVALHO MR, MATIAS MJ, GOFF FE, BASTO MJ, GRACA RC, AIRES-BARROS L, ROCHA L (2008) Origins of high pH mineral waters from ultramafic rocks, Central Portugal. Appl Geoch 23:3278-3289.

MARQUES JM, EGGENKAMP HGM, GRAÇA H, CARREIRA PM, MATIAS MJ, MAYER B, NUNES D (2010) Assessment of recharge and flowpaths in a limestone thermomineral aquifer system using environmental isotope tracers (Central Portugal). Isot Env Health Stud 46:156-165.

MATIAS MJ, MARQUES JM, FIGUEIREDO P, BASTO MJ, ABREU MM, CARREIRA PM, RIBEIRO C, FLAMBO A, FELICIANO J, VICENTE EM (2009) Assessment of pollution risk ascribed to Santa Margarida Military Camp activities (Portugal). Environ Geol 56:1227-1235.

OHGAMI H, TERAO T, SHIOTSUKI I, ISHII N, IWATA N (2009) Lithium levels in drinking water and risk of suicide. British J Psychiat 194:464-465.

PIMENTEL JM (1852) Memoria e estudo chymico da agua mineral de S. João do Deserto em Aljustrel. Lisboa.

QUIROZ JA, MACHADO-VIEIRA R, ZARATE JR CA, MANJI HK (2010) Novel insights into lithium´s mechanisms of action: Neurotrophic and neuroprotective effects. Neuropsychobiol 62:50-60.

RAMALHO EC, LOURENÇO MC (2005) As Águas de Alfama – Memórias do passado da cidade de Lisboa. Rev APRH, 26, 101-112

RASMUSSEN ES, LOMHOLT S, ANDERSEN C, VEJBÆK OV (1998) Aspects of the structural evolution of the Lusitanian Basin in Portugal and the shelf and slope area offshore Portugal. Tectonophysics, 300, 199-225.

RAPANT S, BODIŠ D, VRANA K, CVEKOVÁ V, JORDÍK J, KRĔMOVÁ K, SLANINKA I (2009) Geochemical atlas of Slovakia and examples of its applications to environmental problems. Environ Geol 57:99-110.

REIMANN C, KASHULINA G, DE CARTIAT P, NISKAVAARA H (2001) Multi-element, multi-medium regional geochemistry in the European Arctic: element concentration, variation and correlation. Appl Geochem 16:759-780.

RIBEIRO A, KULLBERG MC, KULLBERG JC, MANUPPELLA G, PHIPPS S (1990) A review of Alpine tectonics in Portugal: foreland detachment in basement and cover rocks. Tectonophysics, 184, 357-366.

RIBEIRO A, MUNHÁ J, DIAS R, MATEUS A, PEREIRA E, RIBEIRO L, FONSECA P, ARAUJO A, OLIVEIRA T, ROMAO J, CHAMINE H, COKE C, PEDRO J (2007) Geodynamic evolution of the SW Europe Variscides. Tectonics, 26, TC6009.

SALMINEN R, BATISTA MJ, BIDOVVEC M, DEMETRIADES A, DE VIVO B, DE VOS W, DURIS M, GILUCIS A, GREGORAUSKIENE V, HALAMIC J, HEITZMANN P, LIMA A, JORDAN G, KLAVER G, KLEIN P, LIS J, LOCUTURA J, MARSINA K, MAZREKU A, O'CONNOR PJ, OLSSON SÅ, OTTESEN R-T, PETERSELL V, PLANT JA, REEDER S, SALPETEUR I, SANDSTRÖM H, SIEWERS U, STEENFELT A, TARVAINEN T (2005) Geochemical Atlas of Europe. Geological Survey of Finland, Espoo.

SCHERMERHORN LJG (1982) Framework and evolution of Hercynian mineralization in the Iberian Meseta. Com Serv Geol Portugal 68:91-140.

SCHMIDT G (2001) Terrestrische Freiluft-Salinen der Iberischen Halbinsel. Der Anschnitt 53:80-87.

SCHRAUZER GN, SHRESTHA KP (1990) Lithium in drinking water and the incidences of crimes, suicides, and arrests related to drug addictions. Biol Trace Elem Res 25:105-113.

SHARMA SK (2003) High fluorine in groundwater cripples life in parts of India. Med Geol Newsl 7:15-16

SMITH DB, REIMANN C (2008) Low-density geochemical mapping and the robustness of geochemcial patterns. Geochem Expl Environ Anal 8:219-227.

THADEU D (1989) Portugal. In: Dunning FW, Garrard P, Haslam HW, Ixer RA eds, Mineral Deposits of Europe, Vol. 4/5: South West and Eastern Europe with Iceland. The Institution of Mining and Metallurgy & The Mineralogical Society of London, 197-220.

UPHOFF TL (2005) Subsalt (pre-Jurassic) exploration play in the northern Lusitanian basin of Portugal. AAPG Bul 89:699-714.

VAN SAMBEEK MHG, EGGENKAMP HGM, VISSERS MJM (2000) The groundwater quality of Aruba, Bonaire and Curaçao: a hydrogeochemical study. Geol Mijnb / Neth J Geosc 79:459-466.

WHO [WORLD HEALTH ORGANIZATION] (2004) Guidelines for drinking-water quality. Vol 1: 3rd edition.

THE AUTHORS

Hans Eggenkamp

Research Professional at Onderzoek & Beleving and Associate Researcher at Utrecht University. Was Investigador Auxiliar at Instituto Superior Técnico, Technical University of Lisbon from 2008 until 2013. Ph.D. in Stable Isotope Geochemistry, with subject "$\delta^{37}Cl$; The geochemistry of chlorine isotopes", obtained at Utrecht University (1994). His main areas of research are isotope geochemistry of chlorine and bromine isotopes, hydrogeochemistry, water-rock interaction, experimental geochemistry and analytical geochemistry. He has participated in research programs in the Netherlands, the United Kingdom, France and Portugal. He is author of the monograph "The Geochemistry of Stable Chlorine and Bromine Isotopes" and is author or co-author on several scientific papers on chlorine and bromine isotope geochemistry and hydrogeochemical papers concerning mineral and thermal waters from Portugal.

José Manuel Marques

Assistant Professor at Instituto Superior Técnico, Technical University of Lisbon. Between 2007 and 2014 he was Coordinator of the Centro de Petrologia e Geoquímica of Instituto Superior Técnico. Ph.D in Mining Engineering, with subject "Geochemistry of low-temperature geothermal fluids and water-rock interaction" obtained at Technical University of Lisbon, Instituto Superior Técnico (1999). His main scientific areas of research are Hydrogeology, Water-rock interaction, Geochemistry of Geofluids and Isotope Hydrology. Other scientific areas of interest are Mineralogy, Petrology, Engineering Geology, Environmental Engineering and Environmental Geology. He has participated in several National and International R&D Projects mainly focused on the geochemistry of surface and groundwaters and water-rock interaction studies, based on chemical and isotopic composition of waters and rocks. He is author and co-author of several scientific papers in Book Chapters and National and International Journals.

Maria Orquídia Neves

Assistant Professor at Instituto Superior Técnico, Technical University of Lisbon. Ph.D in Mining Engineering, with subject "Closed Mines and its Geochemical Environmental Impacts. The case study of Cunha Baixa Uranium Mine (Viseu)", obtained at Technical University of Lisbon, Instituto Superior Técnico (2003). Her main scientific areas of research are Environmental Geochemistry (mining, industrial, agricultural inorganic and organic contamination, risk assessment) and Chemical Analysis. Another scientific area of interest is Medical Geology, dealing with relationships between geological and anthropogenic factors and effects on human and animal health. She has participated in National and International R&D Projects mainly focused on the geochemistry of groundwaters, soil-water-plant interactions and health risks. She is author and co-author of several scientific papers in National and International Journals.